snakes
OF THE EASTERN UNITED STATES

Whit Gibbons

snakes

OF THE EASTERN UNITED STATES

The University of Georgia Press   Athens

A Wormsloe
FOUNDATION
nature book

© 2017 by the University of Georgia Press
Athens, Georgia 30602
www.ugapress.org
All rights reserved
Designed by Mindy Basinger Hill
Set in Quadraat
Printed and bound by Four Colour Print Group
The paper in this book meets the guidelines for permanence and durability of the Committee on Production Guidelines for Book Longevity of the Council on Library Resources.

Most University of Georgia Press titles are available from popular e-book vendors.

Printed in China

21 20 19 18 17 P 5 4 3 2 1

Library of Congress Cataloging-in-Publication Data

Names: Gibbons, Whit, 1939– , author.

Title: Snakes of the Eastern United States / Whit Gibbons.

Description: Athens : The University of Georgia Press, 2017. | Series: Wormsloe Foundation nature book | Includes bibliographical references and index.

Identifiers: LCCN 2016026147 | ISBN 9780820349701 (flexiback : alk. paper)

Subjects: LCSH: Snakes—East (U.S.) | Snakes—East (U.S.)—Identification.

Classification: LCC QL666.O6 D5976 2017 | DDC 597.96—dc23

LC record available at https://lccn.loc.gov/2016026147

CONTENTS

All about Snakes 1
   Why Snakes 3
   General Biology of Snakes 4
   Snake Diversity 9
   Food and Feeding 13
   Predators 20
   Defense 22
   Reproduction 26
   Locomotion 30
   Activity 31
   Temperature Biology 35
   Identifying Snake Species of the Eastern United States 36

Species Accounts 45
   Introduction 47
   How to Use the Species Accounts 48
   Small Terrestrial Snakes 51
      Smooth Earthsnake · 53   Rough Earthsnake · 57   Brownsnake · 61   Red-Bellied Snake · 65   Lined Snake · 69   Kirtland's Snake · 72   Short-Headed Gartersnake · 76   Eastern Wormsnake · 79   Western Wormsnake · 83   Southeastern Crowned Snake · 86   Rim Rock Crowned Snake · 89   Florida Crowned Snake · 92   Flat-Headed Snake · 96   Short-Tailed Snake · 99   Pine Woods Snake · 102   Ring-Necked Snake · 106
   Midsized Terrestrial Snakes · 111
      Scarlet Snake · 113   Rough Green Snake · 117   Smooth Green Snake · 121   Plains Gartersnake · 124   Common Gartersnake · 128   Butler's Gartnersnake · 134   Eastern Ribbonsnake · 137   Western Ribbonsnake · 141   Eastern Hognose Snake · 145   Western Hognose Snake · 150   Southern Hognose Snake · 154   Mole and Prairie Kingsnakes · 158   Milksnake · 162   Scarlet Kingsnake · 167

Large Terrestrial Snakes · 171
  Common Kingsnake · 173   Pine Snake · 179   Louisiana Pine Snake · 184
  Bullsnake · 188   Ratsnake · 192   Corn Snake · 199   Great Plains Ratsnake · 204
  Fox Snake · 208   Racer · 212   Coachwhip · 219   Eastern Indigo Snake · 223
Watersnakes · 229
  Black Swamp Snake · 231   Queen Snake · 235   Graham's Crayfish Snake · 239
  Striped Crayfish Snake · 242   Glossy Crayfish Snake · 245   Northern Watersnake · 249
  Southern Banded Watersnake · 255   Salt Marsh Snake · 260   Plain-Bellied
  Watersnake · 264   Brown Watersnake · 269   Diamondback Watersnake · 273
  Eastern Green Watersnake · 277   Western Green Watersnake · 281   Mud Snake · 284
  Rainbow Snake · 289
Venomous Snakes · 293
  Copperhead · 301   Cottonmouth · 307   Pigmy Rattlesnake · 313   Massasauga · 318
  Timber/Canebrake Rattlesnake · 322   Eastern Diamondback Rattlesnake · 329
  Coral Snake · 335
Introduced Species · 341
  Brahminy Blind Snake · 345   Burmese Python · 349   Northern African Python · 356
  Boa Constrictor · 360

People and Snakes 365
What Is a Herpetologist? 367
Urban Snakes 372
  Most Common Snakes Found in 25 of the Largest Eastern Cities 374
Snakes as Pets 378
Snake Conservation 381
Attitudes about Snakes 387
Frequently Asked Questions about Snakes 388

What Snakes Are Found Where You Live? 391
What Kinds of Snakes Are Found in Your State? 392
Conservation Status of Snakes in the Eastern United States 394
Distribution of All Snake Species 396
Distribution of Venomous Snake Species 397

Glossary 398
Further Reading 402
Acknowledgments 405
Photo Credits 408
Index of Scientific Names 411
Index of Common Names 413

all about snakes

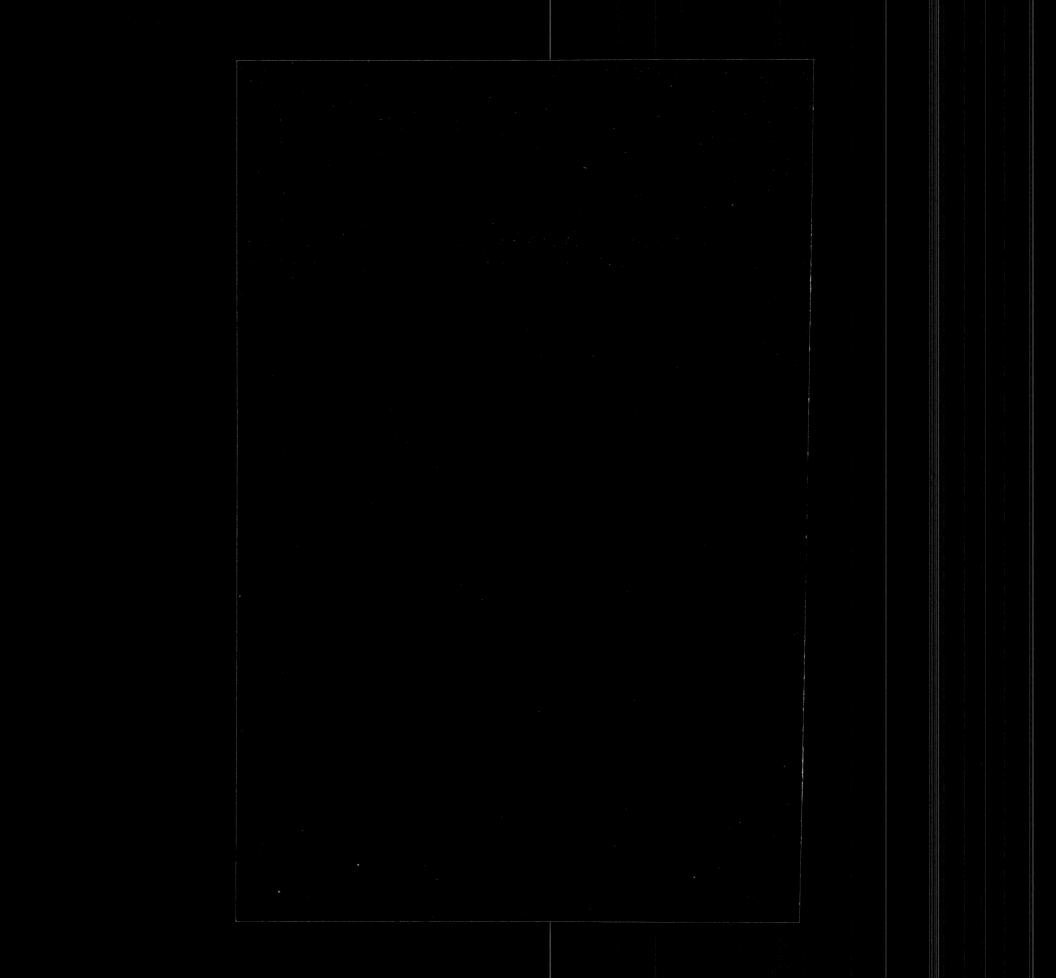

WHY SNAKES?

Nearly everyone is fascinated by snakes. Some fear them, some are attracted to them, but few are indifferent toward them. And although dread of snakes may be one of the most common phobias in the world, today's increased concerns about conservation and environmental issues have led people to pay more attention to the well-being of all wildlife, including snakes.

Most people are interested in learning where snakes live in their region, which are venomous and which are harmless, and how to tell one species from another. This book seeks to teach people about snakes, to help them to identify snake species, and to foster appreciation of snakes as valuable components of our natural heritage. Snakes have been feared, hated, and maltreated for too long, and have been mostly ignored when conservation and environmental issues are under consideration. One goal of this book is to interest young people and adults who may have missed an earlier opportunity to get to know this group of captivating yet often maligned animals. Anyone already familiar with snakes should also learn from some of the recently discovered facts presented about them.

left The eastern hognose snake is geographically widespread and is familiar to many people throughout the eastern states because of its unique behaviors.

below Eastern snakes include a variety of species such as scarlet kingsnakes that are appealing simply because they are colorful.

Most people readily recognize either of the two species of green snakes and know they are inoffensive to humans. Rough green snake (*left*); smooth green snake (*right*).

> **DID YOU KNOW?**
>
> Herpetologists around the world have described more than 3,000 species of snakes.

The eastern United States offers many opportunities for appreciating nature, and my hope is that this book will help to foster excitement and awe when people are fortunate enough to see snakes in the wild or in their own backyard. I encourage people to place the same value on encountering a snake as they do on seeing a dolphin, hearing a frog call, or touching a box turtle. I would like to see everyone develop an acceptance of—better yet, an admiration for—snakes that equals that expressed for many other wild creatures.

GENERAL BIOLOGY OF SNAKES

What is a snake? Snakes are reptiles, just like lizards (their closest relatives), turtles, and alligators. Their most distinctive traits are their elongated body and lack of limbs. These characters impose certain limitations on snakes compared with other animals, yet these same biological features give them unusual abilities as well. Snakes can maneuver through underground burrows and tunnels and negotiate tight passages much better than most other animals of their size. Many are excellent climbers and can move through trees with remarkable agility.

Like most other reptiles, snakes are covered with scales. Even their eyes are protected by clear, transparent scales, which eliminate the need for movable eyelids. The shape, size, and placement of the body scales in relation to each other are different for each species and are commonly used in identification. All snakes shed the outermost layer of skin covering their scales, typically several

Some species of snakes have small geographic ranges that are restricted to the eastern United States. The short-headed gartersnake (*top*) is known from only a few counties in three states (New York, Pennsylvania, and Ohio). The range of the short-tailed snake (*left*) is confined to the north-south sand ridge in central Florida.

Some snakes, such as this red-bellied snake from Sauk County, Wisconsin, have geographic ranges that include all of the eastern states.

Some snakes can exhibit considerable color or pattern variation, even if they are from the same mother. Such variation can sometimes make identification difficult for those unfamiliar with the species, as with the checkered (*top*) and striped (*bottom*) phases of the common gartersnake.

times per year. About a week before shedding their skin, they become rather inactive, their skin color becomes duller, and their eyes become bluish gray. Most snakes shed by loosening the skin on their nose and then literally crawling out of the skin, leaving behind an inside-out remnant. Scientists call this shedding process *ecdysis*.

All snakes have teeth. Some, including seven species found in the eastern United States, have hollow fangs in the front of the mouth that are used to inject venom. Some snakes are rear-fanged, which means they have enlarged, but not hollow, teeth in the back of the mouth. Rear-fanged snakes in the United States are generally harmless to humans. Most snakes have many thin, needlelike teeth in their upper and lower jaws that curve backward, making it difficult for captured prey to escape their grasp. Snakes swallow their prey whole, and they can eat animals much wider than they are themselves because their upper and lower jaws are loosely connected to each other and their bodies can stretch remarkably. Snakes sometimes take more than an hour to swallow very large prey.

The relatively narrow, elongated bodies of snakes require an internal anatomy that differs noticeably from that of most other vertebrates, including other reptiles. Whereas paired organs are usually

DID YOU KNOW?

Many small eastern snakes that are considered harmless to humans have enlarged teeth in the back of the mouth (rear-fanged) that facilitate the transfer of toxic saliva into their prey.

Prior to shedding the skin, a snake's eyes become opaque, as seen in this racer from Ohio. Most snakes remain in hiding during the preshedding period.

A smooth earthsnake from Dawson County, Georgia, begins to shed its skin. Snakes first shed the head and literally crawl out of their skin, leaving it behind, with the tail pointing in the direction they went.

Legless lizards in the genus *Ophisaurus* are frequently misidentified as snakes. Glass lizards, such as this eastern glass lizard from Beaufort County, South Carolina, are readily distinguishable from snakes by having ear openings and movable eyelids, neither of which are found in any snakes.

> **DID YOU KNOW?**
>
> *Some eastern lizards have no legs and are often mistaken for snakes.*

side-by-side in most animals, they tend to be staggered in snakes. For example, one lung is farther down the body than the other, and in most species of snakes only one lung actually functions. Likewise, the kidneys lie one ahead of the other, as do the testes in males and the ovaries in females. The liver is greatly elongated compared with that of a mammal, bird, or reptile of similar size.

The digestive tract of a snake is like those of other vertebrates in having an esophagus, stomach, and small and large intestines that lead to the outside through the *cloaca*, a chamber into which the intestinal, urinary, and genital tracts empty. The cloacal opening, visible on the underside of all snakes and covered by an enlarged scale known as the anal plate (or anal scute), is where the body ends and the tail begins.

The reproductive organs of snakes are also of the same basic design and function as those of other vertebrates, except that male snakes have two copulatory organs (i.e., penises) called *hemipenes* (singular = *hemipenis*). Unless the snake is copulating, the two hemipenes are tucked inside the base of the tail behind the cloaca. Thus, male snakes typically have longer and thicker tails than female snakes. During mating, the male snake fertilizes the female by inserting either his right or left hemipenis into her cloaca. Sperm travels along a groove in the hemipenis and into the female to fertilize her eggs. Some female snakes can store viable sperm in special receptacles for months or even years.

Snakes have the same senses as other vertebrates. Although they cannot typically hear airborne sounds the way mammals do because they have no ear openings and no middle ear, snakes can detect vibrations through the ground or water. Their combined sense of taste and smell is well developed. Snakes flick their forked tongue to gather odor particles from the environment. The tongue transports the particles to a cavity in the roof of the mouth called the *Jacobson's organ* (or *vomeronasal organ*) for chemical identification. Snakes that typically travel above the ground have very good vision and often have enlarged eyes, whereas some burrowing forms have reduced eyes and can discern only shadows. Because snakes have no eyelids, they do not blink and seem always to be staring.

Some snakes, including all pit vipers, have special sense organs in the head region that can detect variation in infrared radiation (temperature). The heat-sensitive organs of pit vipers are in openings called "pits" on the face of the snake between the eye and the nostril. Both species of pythons introduced into

Snakes native to the eastern United States belong to one of three families: Colubridae (nonvenomous species; *bottom left*), Viperidae (the pit vipers; *top left*), or Elapidae (coral snake; *right*).

Florida have similar pits along their upper and lower lips. Even in complete darkness a pit viper or python can detect and pinpoint a subtle change in temperature such as the difference between the body temperature of a mouse and its surroundings. A sense organ that can detect heat is extremely valuable to an animal that hunts at night.

SNAKE DIVERSITY

Although all snakes are superficially alike in lacking limbs, the diversity of body forms, habitats occupied, feeding strategies, and behavior patterns they display is remarkable. Approximately 20–25 families of snakes comprising more than 3,400 species are known worldwide. More than 140 species in 5 families are native to the United States. The eastern United States is the natural home to more than 60 species belonging to 3 families: Elapidae (coral snake), Viperidae (cottonmouth, copperhead, and rattlesnakes), and Colubridae (all remaining native snakes). Four introduced species—Brahminy blind snake, Burmese python, northern African python, and boa constrictor—have become established in some areas of Florida, adding 4 more species and 3 additional families to the total. About 20 native species are *endemic* to the eastern United States, which means that they are not found anywhere else.

Worm snakes are among the smallest snakes native to the eastern United States.

Two or more of six different species of gartersnakes and closely related ribbonsnakes are found in every eastern state. Eastern ribbonsnake (Cape Hatteras, North Carolina; *left*). Eastern gartersnake (Ledges State Park, Iowa; *right*).

Snakes in the eastern United States occupy, or at least enter, virtually every natural habitat, and many thrive in urbanized areas as well. Sixteen species, including the venomous cottonmouth, rely on aquatic habitats for their primary prey of frogs, fish, or aquatic invertebrates such as crayfish. Some species are characteristically found in sandhill habitats; others are most likely to be found in hardwood or pine forests, vegetation along stream margins, rocky outcrops, or open fields. The salt marsh snake is the only species of North American snake that lives permanently in and around brackish water and saltwater.

> **DID YOU KNOW?**
>
> *Approximately 140 distinct species of snakes are found in the United States and Canada. If subspecies are included, the number of different kinds is more than 300.*

Principles of Snake Taxonomy All animal species are given a two-part scientific name that includes a genus name (e.g., *Coluber*) followed by a second name (e.g., *constrictor*), which is referred to as the species *epithet*. Thus, the proper scientific name of the racer is *Coluber constrictor*. Scientific names follow strict rules of zoological nomenclature, with the first person to describe a species using proper methodology being credited for naming it. This convention was first accepted with the publication of *Systema Naturae*, written by Carl Linnaeus in 1758, but not formalized as a standard practice until the late 1800s by the International Commission of Zoological Nomenclature.

Some species have been partitioned into taxonomic units or "races," often called subspecies, that typically occupy geographic ranges that may be contiguous with but separate from other subspecies. A taxonomically recognized subspecies has a third name, the subspecies name, added to the standard two names of the species (e.g., *Coluber constrictor priapus*). Different subspecies of the same species are able to interbreed naturally in areas of geographic contact. This phenomenon may result in a form of intermediate appearance, a consequence of genetic mixing between individuals of different subspecies within a zone where their ranges overlap. These *intergrade* specimens often possess a mixture of traits of the different subspecies involved.

Common Names Common names do not follow an established code of nomenclature like that accepted for scientific names, and there can be great inconsistency in what people call particular species. For example, the snakes known scientifically as *Pantherophis obsoletus* are called "ratsnakes" by many people, "black snakes" in northern parts of their range, "gray ratsnakes" in some areas, and "chicken snakes" in the South. Professional herpetologists primarily use scientific names, and many do not consider an attempt to standardize common names an important issue or worthwhile endeavor. Some professional herpetologists in North America have attempted to develop a system of "standard English names," but I am in agreement with many other herpetologists, who see no compelling reason to insist on strict usage of a common name for any snake as long as it is clear what species is being referred to, especially if the customary terminology varies from one region to another. Forced standardization of common names often detracts from regional cultural heritage and can result in new nontraditional names that confuse laypeople who do use common names on a frequent basis.

My approach has been to maintain a close agreement with common names that have been proposed for standard usage for particular species and subspecies. However, in some instances I use names that I consider more appropriate while also acknowledging the names proposed for standard usage. In regard to the capitalization of common names, I have deferred to the accepted practice of most editors of books and newspapers in capitalizing only the names of plants and animals that include proper nouns (e.g., Louisiana pine snake) and using lower case for all other common names.

Taxonomic Controversies Taxonomy is the scientific field of classification and naming of organisms. Snake taxonomists strive to classify and name snakes in a way that reflects the ancestral relationships among species. Thus, closely related species are placed together within a genus (plural = *genera*), and closely related genera are grouped together within a family. Taxonomy is not a perfect science. Very rare snakes may be difficult to classify because so little is known of their biology. In addition, taxonomists often disagree about the relative importance of different traits in classification, and thus may have different views on the lineage and ancestral relationships (*phylogeny*) of particular species or groups of species. Sometimes these disagreements result in the scientific name of a species being changed. Modern genetics has allowed herpetologists to resolve many of the old taxonomic debates, although in some instances disputes have actually become more heated because of differing interpretations of the findings.

> **DID YOU KNOW?**
>
> Of the more than 60 snake species native to the eastern United States, 16 were described as new to science before the signing of the Declaration of Independence in 1776.

What's in a Name? In this book I have used scientific and common names that are familiar to most herpetologists and lay people alike. Although I give information about taxonomic disputes, I do not try to resolve them. My purpose is to be certain that the reader knows what snake I am talking about, no matter what name the animal is given.

The "What's in a Name?" section at the end of each species account offers insight into the origin of the common and scientific names of snakes found in the eastern United States. I indicate who formally described the species and in most instances provide the Latin or Greek derivations for the scientific names and the rationale for choosing them. For species named in recognition of particular individuals or localities I provide historical background of personalities and places.

I also identify taxonomic controversies in which herpetologists do not agree on what the proper scientific name (or often even the common name) should be. In this book I have chosen to use scientific names that are currently accepted by most herpetologists. However, not all will have the same opinion about what the current usage of the scientific name should be. My intent in this book is to assist the reader in recognizing and appreciating snakes and their biology anywhere in the eastern United States and to provide information about the origins of their names and professional disagreements over those names in order to further that goal.

> **DID YOU KNOW?**
>
> *Carl (aka Carolus) Linnaeus of Sweden is credited with initially describing and naming almost twice as many species of eastern snakes as anyone else. Yet his censorious attitude toward them is summed up in the following quote about reptiles and amphibians: "These foul and loathsome animals are abhorrent because of their cold body, pale color, cartilaginous skeleton, filthy skin, fierce aspect, calculating eye, offensive voice, squalid habitation, and terrible venom; and so their Creator has not exerted his powers to make many of them."*

FOOD AND FEEDING

All snakes are carnivores (that is, they eat animals rather than plants), and all snakes swallow their prey whole. Although some species have very specialized diets, snakes as a group eat animals from almost every major animal group. Small snakes eat insects, spiders, and earthworms; larger ones eat rabbits, birds, and bullfrogs. The ability to overpower and consume other animals without possessing limbs is a rather remarkable feat.

Of course, snakes must first find prey before eating it. Among the senses used by snakes to locate and identify their prey are sight, smell, touch, and in-

above Racers eat a wide variety of prey, including other snakes, by biting and holding them without constriction. Here, a black racer eats a brown watersnake.

left All snakes have forked tongues. The eastern diamondback rattlesnake uses its tongue to track its prey after striking and biting it.

frared heat detection, and many snakes use all or most of these senses. Some species, such as racers and coachwhips, primarily use their particularly good vision to find prey. Many snakes track prey by following its scent. Pit vipers, such as rattlesnakes, use scent to find appropriate ambush locations and then use their vision and infrared detecting abilities to locate prey, often in near total darkness. Rattlesnakes and other pit vipers usually strike, inject venom, and release their prey, which will typically run away and eventually die. They then rely on their remarkable ability to track down the dying animal. Brown watersnakes will sometimes drape themselves over a limb, hang their head and upper body into the water, and grab, with spectacular speed, any fish that touches or gets close to it.

A timber rattlesnake in New York waits in an ambush position for a rodent to run across a log. A large rattlesnake may wait like this for days or even weeks. All pit vipers rely on venom to kill their prey.

A black ratsnake begins to swallow a northern cardinal that it has just killed by constriction.

In addition to capturing live prey, cottonmouths will sometimes scavenge road-killed snakes and frogs.

Herpetologists categorize snakes into two broad categories, ambush or active foragers, based on how they hunt their prey. Ambush predators, such as timber rattlesnakes, may remain motionless for hours or even days waiting for prey to pass within striking distance. Juveniles of many species of pit vipers, including copperheads, cottonmouths, and pigmy rattlesnakes, sit very still and wiggle a brightly colored tail tip to attract potential prey, such as frogs or lizards, within striking distance. Wide-ranging, active foragers, such as racers and coachwhips, can cover large areas in search of prey. Some snakes use both strategies, ambush and active foraging, to find prey.

Snakes can also be categorized by the type of food they eat. *Generalists* eat a wide variety of animals, sometimes even scavenging recently dead ones; specialists focus on one or a few prey types. Racers and kingsnakes are good examples of generalists because they eat many types of animals—including birds, mammals, reptiles, and even other snakes. Hognose snakes, which eat primarily toads, are considered dietary specialists. *Specialists* often have unique adaptations that help them to find, capture, and subdue their prey. For example, mud snakes, which feed almost exclusively on large salamanders, have a spine on the tip of their tail that is thought to help them anchor themselves while holding on to their slippery prey.

Snakes have several techniques for subduing prey. The most straightforward approach is to bite the animal to be eaten and hold on while gradually swal-

Northern watersnakes have been documented to eat more species of fishes (80) and amphibians (30) than any other U.S. snake.

A large lump is usually evident after snakes eat a large meal, as seen in this northern watersnake after eating a fish.

An eastern kingsnake in Virginia kills an eastern gartersnake by constriction. Kingsnakes, ratsnakes, and pine snakes are powerful constrictors.

lowing it alive. All snakes native to the eastern United States have numerous sharp, backward-pointing teeth that direct the prey in one direction—toward the stomach. The indigo snake has a powerful bite that crushes soft-bodied prey and will often hold down its primary prey, other snakes (even rattlesnakes), with its large body and chew on the victim's head until dead. Constriction is another method widely used to kill living prey. After striking and biting its prey, the snake loops its body once or more around the prey and tightens its coils enough to kill the prey. Smaller constricting snakes generally suffocate their prey, but larger, stronger snakes may actually squeeze tight enough to cause cardiac arrest. It makes sense for constricting snakes to squeeze their prey as tight as they can and incapacitate it as quickly as possible because many snakes feed on animals that might cause them harm by biting or scratching during the process of constriction. Quite a few snakes—more than you might think—use venom to subdue or kill their prey. The saliva of many species can paralyze or even kill small animals when the venom enters wounds made by the sharp teeth. But only seven species in the eastern United States—six pit vipers and the coral snake—actually inject venom through hollow fangs.

Some of the specialist feeders have unusual structural, physiological, and behavioral strategies that allow them to capture and eat their chosen prey. Crowned snakes eat centipedes, which are venomous. The crowned snake bites the centipede behind the head in a way that prevents the centipede from biting back. The snake forces its enlarged rear teeth into the centipede's body and holds on tightly until its venomous saliva has paralyzed the centipede. The glossy crayfish snake and striped crayfish snake use their enlarged, chisel-like teeth to grasp and consume hard-shelled crayfish. Virtually all snakes can go long periods without eating—many days, weeks, or in some instances months—

right An eastern hognose snake finishes its meal of a southern toad. Hognose snakes will eat other prey, but their primary diet consists of toads.

below Among eastern snakes, queen snakes have one of the most specialized diets—soft-bodied crayfish in the process of molting.

and in cooler parts of the eastern United States, most do not eat at all during the winter. Likewise, by remaining dormant underground, most snakes are able to survive periods of drought when food may be scarce.

PREDATORS

Snakes and snake eggs are eaten by an enormous variety of animals throughout the eastern United States. Almost any large animal that eats other animals will eat a snake if it can. The most vulnerable snakes are the small ones, which must fear small creatures such as spiders, toads, and shrews in addition to larger predators. Large carnivorous mammals such as bobcats and coyotes, as well as medium-sized ones such as raccoons, skunks, and opossums, are potential threats to larger snakes, although the snake's size can become a deterrent in some instances. Otter and mink readily prey on snakes in wetland habitats. Birds of prey, especially hawks, capture snakes they find crossing open areas and lying on limbs along riverbanks. Snakes that travel overland at night fall victim to owls; small woodland snakes become prey for ground-scratching birds; and aquatic snakes along wetland margins become the quarry of wading birds. Domestic and feral cats kill thousands of snakes each year in residential areas.

Large fish such as gar, catfish, and bass are an ever-present hazard for aquatic snakes, and alligators are a constant menace to all snakes in the water, even large cottonmouths. Some amphibians eat snakes as well. Bullfrogs and the giant salamanders known as amphiumas are common predators of small to medium-sized aquatic snakes.

> **DID YOU KNOW?**
>
> Watersnakes that feed on catfish are occasionally found with the fish's spines sticking out through their body walls. The snakes generally recover without complications.

Eastern snakes have a wide variety of mammalian predators, including raccoons, feral pigs, and coyotes.

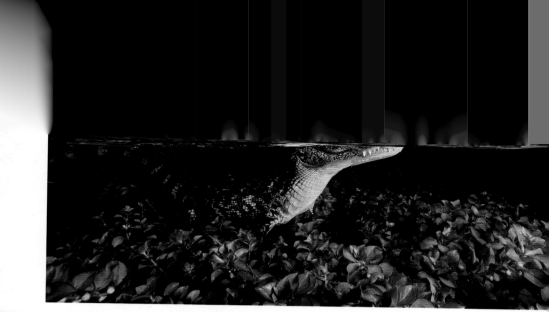

The American alligator is a major natural predator of watersnakes and cottonmouths, which often swim on the surface at night.

left Hawks and owls are common predators of snakes throughout the country. Red-shouldered hawks are frequent predators on snakes living in river swamps and other aquatic habitats in the eastern United States.

right Domestic cats that are mellow house pets can become deadly predators for small snakes if allowed outside in suburban areas.

Although many predators will eat snakes when given the opportunity, only one type of predator in the eastern United States specializes on them—other snakes! Because snakes are long and thin and must swallow their prey whole, other snakes make ideal prey for many species. That is, it is much easier to swallow a large prey animal that is long and slender instead of round or oval-shaped. Typical snake predators characteristic of the Southeast include coral snakes, coachwhips, and indigo snakes, but many also abound in the Northeast and Midwest, including common kingsnakes, racers, and copperheads. Some snake-eating snakes sometimes swallow snakes larger than themselves. One

The eastern indigo snake is one of the most dangerous predators for other snakes. Here an indigo is consuming a large coachwhip.

DID YOU KNOW?

A native scorpion has been suggested to be a natural predator of the rim rock crowned snake.

reason snake-eating snakes can effectively target other snakes is because they can pursue them in their favorite hiding places: underground burrows, beneath logs, and in rock crevices.

Snake eggs are also vulnerable to a wide variety of animals, especially mammals but also other snakes and even insects. Scarlet snakes specialize on reptile eggs, and some herpetologists consider nonnative fire ants to be a major threat to the eggs of species such as the southern hognose snake.

DEFENSE

Snakes in the eastern United States exhibit remarkable—and sometimes very entertaining—responses to threats from predators and people. The first level of defense for most snakes anywhere is to go unseen, which they achieve by hiding out of sight or being effectively camouflaged. Once discovered, the initial response of most species, including venomous ones, is to flee to safety if an escape route is readily available.

When a clean escape seems unlikely, a snake may try to fool or distract a predator. Many try to bluff their way out of the situation. The most common approach, used by many species, is to make the head and body look bigger. Some harmless watersnakes flatten and expand their heads and even their entire bodies so that they resemble a broad-headed, thick-bodied cottonmouth. Thus, venomous snakes often cannot be distinguished from harmless species simply because they have large, "arrow-shaped" heads.

When confronted or captured, many snakes release feces or foul-smelling musk from glands in their cloaca. Each species of snake has its own odor, ranging from the sickeningly sweet, almost perfumed scent of gartersnakes to the nauseatingly thick and overpowering musk of large watersnakes. Most people consider ratsnakes and ring-necked snakes to have a particularly foul musk. Some species, such as eastern diamondback rattlesnakes, will even spray their musk a short distance if captured. The musk released by snakes presumably discourages some predators that would prefer not to eat something that smells (and likely tastes) so bad. People familiar with the distinctive musk smells of

The open-mouthed defensive display is the trademark of a cottonmouth when threatened and unable to readily escape.

Even venomous timber rattlesnakes attempt to hide from potential predators, including humans.

left An eastern hognose snake from Wisconsin spreads its neck in the characteristic defensive display of the species.

right The closing act of an eastern hognose snake's defensive performance is to "play dead" by rolling over on its back with its mouth open. It may even bleed from the mouth and if it has recently eaten a toad will regurgitate it.

different species can often detect the presence of a snake that feels threatened but has not yet been seen.

Other defensive displays include the red-bellied snake's curling of its upper lips to reveal what appear to be large teeth; the pine snake's openmouthed loud hissing; the cottonmouth's gaping; and the death feigning of several species, with hognose snakes giving the most spectacular (and realistic) performance, which typically follows hissing and spreading of the neck region.

Mud snakes, ring-necked snakes, and coral snakes, all of which have brightly colored undersides, will sometimes put their head under their body, curl their tail, and turn it upside down, exposing the brightly colored undersurface. Whether the display is a threat or an attempt to divert the attack away from the head and toward the tail is unknown.

Some defensive behaviors are used in response to specific predators. For example, kingsnakes are immune to the venom of pit vipers. Thus, rather than trying to defend itself by biting, a pit viper attacked by a kingsnake will arch its back toward its attacker, apparently aware that its typical biting defense would be ineffective.

When feeling threatened and unable to

> **DID YOU KNOW?**
>
> Pine snakes make a hissing sound that resembles the whirring sound of a large, rattling rattlesnake.

24 *All about Snakes*

The defensive rattling of a large canebrake rattlesnake with a long string of rattles can be heard several yards away and serves as a warning to any potential predator, deer, or human.

A mud snake will not bite a person defensively but may raise its tail to resemble its head in an effort to divert a predator's attack.

Many nonvenomous snakes will open their mouth in a threatening pose when frightened and may bite. The saliva of common gartersnakes can be mildly toxic and cause swelling and itching in some people.

The eastern indigo snake spreads its neck vertically and hisses when threatened.

Some species of snakes, including eastern indigo snakes and cottonmouths, have male-male combat during the mating season, a rarely seen behavior.

DID YOU KNOW?

In snake species that have male-male combat, males usually get larger than females.

escape, many snakes will bite an attacker, although many species will not bite a person under any circumstances. A bite can have serious consequences for people or predators if the snake is venomous. Bites from the larger watersnakes, racers, and ratsnakes can sometimes draw quite a bit of blood, but typically the injury is equivalent to a scratch and requires no more than simple cleaning with soap and water, and possibly the application of an antiseptic. A bite from a pit viper can be painful and even dangerous depending on the species of snake, the location of the bite, and the amount of venom injected. However, even for venomous snakes, biting is usually a last resort, and in many instances only a small amount or no venom is injected.

REPRODUCTION

Snakes are similar to mammals and birds in their general reproductive patterns. Males mate with females by means of internal fertilization, which is sometimes preceded by elaborate courtship behaviors that may include combat between adult males. Male-male combat among some of the large species of snakes (e.g., ratsnakes, kingsnakes, cottonmouths, and rattlesnakes) is a rarely seen but apparently widespread behavior. During combat, two males of the same species face each other, lift their heads above the ground, and entwine the front part of their bodies. The snake that can force its opponent to the ground generally wins the bout. Biting or injury is apparently rare in these combat encounters, even among venomous species. Because the larger male generally wins and gets to mate with the female, males of species with male-male combat rituals are typically larger than females. In species that do not have male-male combat, such as watersnakes and gartersnakes, the females are usually appreciably larger than the males.

As is the case with male-male combat, herpetologists have not observed the courtship of most species of snakes in the wild, but each species is presumed to have a distinctive ritualistic activity. Often the male crawls on top of or beside the female, ultimately placing his cloaca adjacent to hers. During courtship, snakes often twitch or jerk erratically, and in some species the male bites the female on the neck or head. The female generally signals her willingness to mate by lifting her tail and allowing copulation to occur. Gartersnakes and some watersnakes are noted for courtship behavior in which two or several males attempt to mate with a single female at once or in succession.

Many—possibly most—snakes use chemical signals called *pheromones* as part of their mating behavior. Generally, the female snake releases a pheromone as a signal to males that she is ready to mate. Male snakes may follow

Snake courtship and mating are seldom seen in the wild. In most species, such as the coachwhip (*above*), the male finds females ready to mate by following pheromone trails. Male timber rattlesnakes (*left*) use erratic body movements and tongue flicks to entice the female.

Female mud snakes lay eggs in late spring or early summer and will stay with them until they hatch in late summer or early fall.

Brownsnakes give live birth in late summer or early fall.

the pheromones of female snakes long distances overland in seeking an opportunity to mate.

Most species of snakes in the eastern United States mate in the spring, although some mate primarily in late summer or autumn, and others may mate in both seasons. After mating, the female produces young either by laying eggs or by giving birth to live babies. Slightly more than half of the snakes found in the eastern United States lay eggs, and the rest give birth to live young. Research has shown that some of the species (e.g., brownsnakes, Graham's crayfish snakes, and northern watersnakes) that carry the young internally until birth actually nourish the developing babies through placenta-like structures. Future research on live-bearing snakes may reveal that this phenomenon is much more widespread than currently believed. Most egg-laying snakes lay eggs in late spring or early summer, and the eggs hatch in mid-to-late summer or early fall. Live-bearing snakes generally give birth during the summer or early fall. Consequently, many baby snakes are seen by people each year during late August and September.

Egg-laying snakes deposit groups, or clutches, of elongate, white or cream-colored eggs with leathery shells. Nesting sites are usually underground or beneath logs or rocks. Pine snakes tunnel several feet beneath the surface in sandy areas to create a cavity for their eggs. Ratsnakes will often lay eggs under rotting vegetation in the hollow of a tree. The incubation period ranges from only a few days or weeks for some species to more than 2 or 3 months for others. Temperature, and to a lesser degree humidity, affects the success and rate of hatching—warmer temperatures generally increase development rates and thus result in earlier hatching dates. Baby snakes have a tiny egg tooth on the tip of the snout that allows them to slice through their leathery eggshell. The egg tooth is lost the first time the young snake sheds its skin. The females of some snakes, such as mud snakes and rainbow snakes, are believed to remain with the eggs until they hatch. Some pit vipers, all of which are livebearers in

left Baby corn snakes hatching from their eggs

below This baby copperhead from Baker County, Georgia, has shed its skin for the first time and is ready for its first meal.

North America, stay with and presumably protect their newborn young for several days or even weeks after birth. Some species may exhibit complex parental behavior, but it has rarely been seen.

LOCOMOTION

How do snakes move about without limbs? In fact, they get around quite well. The coordinated operation of muscles, flexible ribs, and overlapping belly scutes (the scales on the snake's belly) allows snakes to maneuver down tunnels, over the ground, up trees, and in the water. The coordination of ribs and belly scutes is especially important in allowing a snake to push itself along the ground or a tree limb. Each belly plate is associated with the ends of a pair of ribs. The most common method eastern snakes use to move forward is called *lateral undulation*. As the long muscles down the body of the snake are contracted first on one side and then on the other, the ribs and belly plates push backward against the ground and the snake moves forward. The alternation of contractions on the two sides of the body is so well synchronized that snakes appear to glide effortlessly over the ground, across bushes, or into holes. Swimming snakes undulate their entire bodies to push themselves forward in the water.

Some snakes in the eastern United States occasionally use other forms of lo-

The belly plates (ventral scales) visible on this corn snake coordinate with the ribs and help it to push against a rough surface to crawl forward or climb upward.

comotion more common in snakes from other regions. Racers will sometimes use a *sidewinding* motion to cross a hot highway, for example, the same way sidewinder rattlesnakes move across sandy deserts. A sidewinding snake lets only the front and back ends of its body touch the ground, pushes itself up vertically, and then moves its body horizontally. Large rattlesnakes and other stout, heavy-bodied snakes often use *rectilinear locomotion* when crawling overland or across highways. In this form of movement, muscles on both sides of the body contract simultaneously so that the snake slowly moves forward in a straight line. Snakes using *concertina locomotion* extend the front part of the body forward, anchor it, and then pull the hind end forward—sort of like an inchworm. Most snakes use concertina locomotion when in burrows, and ratsnakes typically use it when climbing. The absence of legs does not make snakes less efficient at locomotion than other animals. The amount of energy a snake uses to move a given distance is similar to that of a lizard that goes the same distance using its legs. Despite their efficiency in traveling over the ground, climbing, and swimming, snakes do not move as rapidly as they sometimes appear to. Black racers, for example, are among the fastest snakes in the eastern United States, but their top speed is still less than 10 miles per hour. No snake can move over the ground faster than the average person can run.

A banded watersnake travels across the flooded margin of a field in North Carolina. All snakes can swim by moving their body in a sinuous pattern.

ACTIVITY

Seasonal Activity Our impression of when snakes are active is influenced by those we see most often. Species that travel overland to get from one wetland habitat to another, to reach hibernation sites, or to search for prey or for mates are the ones people are most likely to see. Snakes and other reptiles become inactive during extremely cold and hot periods because they cannot maintain their body temperatures at a suitable level. Thus, they generally become dormant during cold winters (*hibernation*; some scientists refer to this as *brumation*) and hot, dry summers (*aestivation*). Because most of the southern United States experiences moderate temperatures in most years and most seasons, some snakes are

above Some snakes, like the copperhead, are active during the day in the spring and fall and tend to be more nocturnal during summer.

left Pine snakes are encountered most frequently in the spring, which is their breeding season.

active to some degree during every month in most areas of the region. In more northern latitudes or at higher elevations, snakes in the eastern United States are dormant for longer periods during the winter and are more likely to be active throughout most of the summer. Watersnakes and cottonmouths may aestivate during the hottest and driest parts of the summer when their prey is scarce.

Spring and early summer is generally the period of greatest activity for snakes in the eastern United States, both because many species breed in the spring and also because they are seeking meals as they emerge from hibernation. However, more snakes are actually present and likely to be seen aboveground during late summer and fall because that is when snakes' eggs hatch and live-bearing snakes give birth. The young of some species such as ratsnakes, racers, and gartersnakes are especially common during early autumn. Most of these young snakes are eaten by predators and do not survive until the following spring.

The timing of courtship and breeding influences the aboveground activity of snakes during some seasons. Thus, pine snakes are encountered most frequently in the spring, presumably because that is their breeding season and they are actively pursuing mates. In the spring and fall, snakes are often seen crossing highways, although many do not survive the trip. Canebrake rattlesnakes, especially large males, are commonly killed crossing roads during their late summer and fall breeding season.

Some of the larger species of snakes are more often seen in the fall and spring as they enter and leave their winter hibernation sites. In the mountainous regions, some species gather in large numbers to hibernate in dens, which are usually in rocky outcrops on a sunny southwest-facing slope. Rattlesnakes, copperheads, and ratsnakes frequently den together, and some individuals travel miles to reach a particular denning site. Gartersnakes are noted for hibernating together in large numbers in some regions. Snakes that congregate in this manner generally return to the same dens year after year. In places where snakes must cross a road to reach a hibernation site, numerous snakes may be killed on the road each year, both in the fall and again in the spring as they disperse.

Regional temperature patterns differ from year to year, and snake species respond to these annual variations in different ways. In more southern latitudes, some snakes that disappear underground and become dormant during extended periods of winter cold will remain active during more moderate winters, occasionally appearing aboveground. Cottonmouths will sometimes emerge from underground hibernating sites to warm up in the winter sun, even on relatively cold days. In contrast, canebrake rattlesnakes living in the same region do not typically emerge from hibernation until mid-to-late spring. Likewise, some snakes become active earlier in the year during a warm spring, while others do not emerge from their winter dormancy until late spring regardless of the spring temperatures in a particular year.

Daily Activity Snakes can be active during both day and night in the eastern United States, but daily activity periods vary from one species to the next and with latitude and elevation. In general, snakes tend to be more active during the day in more northern or higher elevation areas and more active at night in the southern states. Some snakes—pine snakes and racers, for example—are

The aboveground activity of some eastern snakes is restricted to nighttime (scarlet snake; *left*), whereas others are active only during the day (eastern coachwhip; *above*).

almost exclusively *diurnal*; that is, they are characteristically active aboveground only during the daylight hours. Scarlet snakes, on the other hand, are *nocturnal*; that is, they come out onto the surface only at night. Some species, notably corn snakes and copperheads, are typically active during the day in the cool seasons and at night during summer. Many watersnakes are active around water at night when searching for nocturnal prey such as frogs but may also be active during the day.

Underground is one of the safest places for a snake, and many species spend most of their lives in root tunnels or beneath logs, rocks, and ground litter. All snakes in the eastern United States are adept at using underground pathways, and some, such as pine snakes and hognose snakes, make their own burrows in sandy or other loose soils. A behavior that tends to make snakes highly visible is basking in the sun, which snakes do to warm themselves. Watersnakes are especially noted for basking on limbs or bushes above water, into which they quickly retreat if disturbed. Likewise, they will often rest on shaded limbs during warm days. Many terrestrial snakes bask on the ground or on rocks during cool periods, often close to a hole or other retreat. Timber rattlesnakes will often bask in large numbers around their hibernation site in the spring before they disperse out into the landscape.

TEMPERATURE BIOLOGY

Temperature affects nearly every aspect of the lives of snakes. It can influence a snake's growth, capture of prey, and ability to escape from predators. A snake's body temperature is determined primarily by the environmental conditions that surround it, and thus most people refer to them as *cold-blooded*, although their blood is not necessarily cold. Scientists refer to cold-blooded animals as *ectotherms*, but both terms mean the same thing.

All snakes, such as the massasauga, have a low-energy lifestyle and may need to eat only a few meals each year.

Most snakes prefer to maintain a body temperature around 86° F (30° C). Humans, other mammals, and birds regulate their body temperature by using internal body heat, which they produce during metabolism. Snakes, however, regulate their body temperature by using different behaviors—such as basking on cool, sunny days and seeking shelter and shade on hot days. Snakes can be particularly susceptible to heat stress.

Low-Energy Lifestyle Just as there are certain advantages to being warm-blooded, or *endothermic*, there are also advantages to being cold-blooded, or ectothermic. Ectotherms have low metabolic rates and thus generally require only about one-tenth the food needed by a similar-sized mammal. This low-energy lifestyle allows snakes to go for long periods between meals and to specialize on food that is available at only certain times of the year. A good example is the egg-eating scarlet snake, which feeds primarily on lizard eggs during late spring and summer. Because snakes do not use metabolic heat to maintain a high body temperature, like mammals and birds do,

Coachwhips prefer higher temperatures than most other eastern snakes and are usually active in the open on warm days.

Watersnakes, such as this brown watersnake from Georgia, typically bask on limbs overhanging the water, alternating between sun and shade depending upon the temperature. They position themselves to permit a rapid escape into the water below when threatened.

they can use much more of the energy from their food for growth and reproduction. Their energetic efficiency also allows snakes to sometimes occur in high densities because each individual snake requires a relatively small amount of food.

IDENTIFYING SNAKE SPECIES OF THE EASTERN UNITED STATES

Being able to identify animals and plants is one of the first steps in developing an appreciation for nature. When people first see a snake, they often ask (or wonder), "What kind is it?" When you know what species a snake is, you can find out what it might eat, how large it can get, and whether it is venomous or harmless. Most of the more than 60 species of snakes native to the eastern United States can be identified using a few key characters, especially when coupled with the geographic location (e.g., a snake found in New Hampshire with a bright green body is without question a smooth green snake, and a shiny black snake with red stripes down its back and sides from coastal Virginia or southern Mississippi will always be a rainbow snake). Such information is fun and easy to learn and can greatly add to your enjoyment of nature in general and snakes in particular.

> **DID YOU KNOW?**
>
> In many species of snakes, females get much larger than males.

Body scale counts and head scale configurations are among the key characters used to tell snake species and subspecies apart, but these are subtle features used primarily by professional herpetologists. Other traits are obvious to even the most casual observer. For example, only four snake species east of the Mississippi River have rattles at the end of the tail. The following guide to the species presents a combination of characters, some quite obvious and others requiring closer examination, that will help distinguish each species from all or most of the others. A few species will never be easily told apart without careful examination, even by an expert in herpetology. And even professional herpetologists will sometimes have difficulty confirming whether the small brown snake in their hand is a smooth or rough earthsnake unless they are holding a magnifying glass. But most snakes inhabiting the eastern United States can be correctly identified with minimal effort by using this book.

Rainbow snakes have an unmistakable color pattern both above and below.

Colors and Patterns Snakes in the eastern United States exhibit an amazing diversity of patterns, ranging from uniform coloration to blotches, spots, and bands or rings to longitudinal stripes. The most common colors are blacks, browns, and grays, which help to offer protection via camouflage. More than 20 of our snake species, however, are brightly colored in shades of red, orange, yellow, or green. The color and pattern of some species—such as rainbow snakes, both species of green snake, and southern hognose snakes—rarely vary and thus, can be used with confidence in identification throughout the eastern United States (see the individual species accounts for descriptions). Other species, such as gartersnakes, red-bellied snakes, and eastern hognose snakes, exhibit variation—even at the same locality—that makes color and pattern un-

The glossy crayfish snake has a distinctive pattern of two rows of dark spots on a light belly that is useful for identification.

Identifying Snake Species 37

reliable or at best confusing as identifying characters. In some species, intergrades that possess color patterns and scale counts intermediate between two or more subspecies may occur where the geographic ranges of the subspecies overlap. All species can exhibit aberrant color patterns on rare occasions (e.g., solid black copperheads and albino rattlesnakes) that can make them difficult to identify. In a few species, in particular the ratsnake, coachwhip, and racer, the juveniles may look entirely different from the adult, but photographs of adult and juvenile individuals from each species are given in the species accounts to aid in identification.

Size as a Character in Identification Eastern snakes range in body size from the tiny wormsnakes and earthsnakes to the enormous pine snakes and indigo snakes, which may be nearly ten times as long and weigh a thousand times more than the smaller species. Because it is a relatively easy measurement, body length is the standard method of determining a snake's size, although body shape and bulk can also be important factors for identification. A 2-foot-long southern hognose snake, for example, will weigh much more than a ribbonsnake of the same length. It is important to remember that every snake will change greatly in length during its lifetime (many will be five times longer as adults than at

The tail of a ribbonsnake is often more than a third of the total body length.

38 *All about Snakes*

The belly scales of all snakes native to the eastern states are in a single row, but only in pit vipers are the scales below the tail (subcaudals) usually in a single row. All subcaudals of nonvenomous snakes and coral snakes are usually in a double row. The configuration of the single or double rows of subcaudals is one way to determine if badly damaged road-killed specimens are venomous, as the tail is often the last part of the body to deteriorate. The pattern of scales is also recognizable on shed skins.

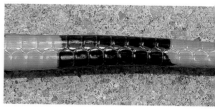

birth), and many species vary in body proportion based on their sex, reproductive condition (for females), and recent feeding success. Therefore, body shape and bulk is not always a reliable character. It may be worthless for differentiating between some species and highly reliable for separating others. For example, a baby coachwhip or pine snake is longer just after hatching than a crowned snake or pine woods snake will ever be.

Another size trait sometimes useful for distinguishing among species is the length of the tail in relationship to the total body length. For example, rattlesnakes have very short tails relative to their body length (less than 10 percent), while the tail of ribbonsnakes can be more than a third of the total body length. In all eastern snakes, the male's tail, from birth through adulthood, is always proportionately longer than a female's of the same species.

How Can You Tell If a Snake Is Harmless or Venomous? People probably ask more questions about the identification, behavior, and ecology of snakes than about any other group of animals in the eastern United States. One of the first questions a person wants to know on finding a live or dead snake, or even

> **NOTE**
>
> Some of the characters used to distinguish venomous snakes may be difficult to see from a safe distance. The best way to identify whether a snake belongs to a venomous species is to learn what each of the venomous and nonvenomous species in your region looks like.

The elliptical pupil and the heat-sensitive pit between the eye and nostril, characteristic of all pit vipers, are visible on this copperhead from Brown County, Indiana.

a shed skin, is whether the snake is venomous. All seven of the venomous species found in the eastern United States differ from all of the harmless species by having fangs in the front part of the mouth, although it is not usually necessary (or recommended) to open a snake's mouth to determine its identity. All six of the pit vipers (rattlesnakes, cottonmouth, and copperhead) have three distinctive characteristics that no other snake in the region has: (1) they have an opening (the heat-sensitive pit) in the side of the head between the nostril and the eye; (2) most of the scales on the underside of the tail are in a single row rather than in a double row (see photos); and (3) the pupils in the eyes of pit vipers appear elliptical in most situations, whereas all other eastern snakes (including coral snakes) have distinctly round pupils.

In the southeastern United States, where venomous coral snakes and several coral snake mimics occur, many children and adults are taught a simple rule that readily separates an eastern coral snake from all of the harmless snakes of the region that might look similar: "Red touch yellow, kill a fellow—red touch black, friend of Jack" (or "venom lack"). That is, any southeastern snake with brightly colored red, yellow, and black rings in which the wider red rings

A shiny black snout followed by broad yellow and black bands distinguish the coral snake from any other eastern U.S. snake.

touch the narrower yellow ones is almost certainly a coral snake. The outlines of the rings are visible even on a shed skin.

One common misconception that can cause a nonvenomous snake to be misidentified as venomous is the erroneous belief that only rattlesnakes vibrate their tails. Many harmless snakes, especially large terrestrial species such as racers, pine snakes, and kingsnakes, vibrate the tail when confronted, and in dry leaves may actually sound like a rattlesnake. Also, a large, triangular head does not necessarily identify a snake as venomous. Many of the large but harmless watersnakes flatten and expand the head when alarmed, making them look very similar to the venomous cottonmouth.

Key Traits to Look for in Identifying Snakes A few characteristics can be used to narrow the field by excluding certain species.

Whether a snake's body scales on the back and sides are keeled with a noticeable ridge down the center of each body scale like the timber rattlesnake (*top*) or smooth with no ridge like the racer (*bottom*) can be useful in species identification.

SCALE TYPE SMOOTH OR KEELED (*see photos*) One of the most reliable traits for distinguishing most species is whether the body scales are mostly keeled (i.e., with a ridge down the center that makes them appear rough) or mostly smooth. For example, all of the pit vipers and most of the watersnakes have keeled scales, and more than 20 of the harmless snake species and the coral snake have mostly smooth scales. A few species, such as ratsnakes and corn snakes, have weakly keeled scales, and the males of some smooth-scaled species, such as indigo snakes and striped crayfish snakes, have a few keeled scales at the base of the tail; indigo snakes may even have weakly keeled scales on the back. The imprint of the keel, when present, is visible on each scale of a shed skin, too.

ANAL PLATE In most snake species, the last belly scale, which is called the anal scute or anal plate and which covers the cloaca and precedes the tail scales, is either whole and undivided or divided into two scales. The trait can be especially useful when you are identifying a shed skin because it can help you narrow down possible species.

BODY SHAPE Snakes vary in shape from species that are characteristically robust for their length to those that are usually very slender. Like other biological traits, shape can vary considerably within a species (e.g., although cottonmouths are typically heavy bodied, thin individuals are occasionally encountered), but most species are relatively consistent in their overall body shape.

PATTERN AND COLOR The presence or absence of encircling rings, bands, blotches, or longitudinal stripes can be used to distinguish many species, although in some species appearance can vary significantly from one individual to the next. For example, gartersnakes usually have three yellow stripes, but in some individuals a checkered pattern predominates and the stripes are less distinct. Also, some watersnakes, cottonmouths, and eastern hognose snakes turn darker with age, so that the body markings of larger individuals may not be visible. Likewise, the young of some of the larger species look very different from the adults.

Cottonmouths are typically heavy bodied.

DISTINCTIVE CHARACTERS Special traits or characteristics can often be useful in narrowing the field to only a few possibilities. Behavior can be an instant clue to a snake's identity; for example, the open-mouthed gape of a cottonmouth, the rattling of a rattlesnake (not to be confused with the tail vibration of harmless species), or the cobralike display of a hognose snake. A large black and red snake with a hardened, spiny tail tip will be a mud snake or rainbow snake. A distinctive color pattern of red, yellow, and black rings should serve immediate notice that the snake is a coral snake (see page 335) or one of the harmless mimics (scarlet kingsnake, scarlet snake, or Louisiana milksnake). Knowing the distinctive behaviors, appearances, and structural characteristics of the different species can be a useful tool in identifying snakes in the wild.

GEOGRAPHIC LOCATION Where a snake is found in the wild often provides an instant clue about its identity. Many species native to the eastern United States

This racer has just captured a leopard frog, which it will swallow alive.

Northern watersnakes are rarely found far from water.

are not found in northern states where it is colder, and a few northern species, such as Butler's gartersnake and massasaugas, are not found in southern states. Copperheads and canebrake rattlesnakes are absent from the southern half of Florida; diamondback and pigmy rattlesnakes are absent from Virginia. Identification of a local species can often be narrowed considerably by eliminating species that do not naturally occur in the region. The table on page 392 at the back of this book will allow you to identify exactly which species occur in each state.

HABITAT The habitat where a snake is found can be another important clue to its identity. Although exceptions always exist, most watersnakes are unlikely to be found far from water, and typical sandhills species such as southern hognose snakes and pine snakes are unlikely to be found near water. Even the specific location within the habitat may provide meaningful information. For example, a coral snake or ring-necked snake would not be expected to be in a bush or tree.

TIME OF DAY The active period of a species is often restricted to daytime or nighttime. A snake found crawling around in the day is highly unlikely to be a scarlet snake, and a snake found crossing a road at midnight is typically not going to be an eastern or southern hognose snake.

All about Snakes

species accounts

INTRODUCTION

The following species accounts are designed to help the reader become familiar with every species of snake native to the eastern United States as well as with four introduced species that have become established, reproducing residents in Florida. The accounts are grouped into six categories based on body size and ecology rather than arranged in the traditional taxonomic groupings seen in most field guides. The six categories are (1) small, (2) midsized, and (3) large terrestrial snakes; (4) watersnakes (i.e., snakes associated with aquatic habitats); (5) venomous snakes (i.e., snakes that use venom to kill their prey and are potentially dangerous to humans); and (6) introduced species (i.e., species that are not native to the United States but that have been brought here by humans and are now established). Although these groupings are somewhat subjective, and the size ranges of juvenile midsized snakes often overlap those of small snakes, they are easily understandable and provide an approach that will help people to become aware of the similarities and differences among the snakes in the eastern United States.

A species' geographic range is indicated on the map that accompanies each account. A smaller map shows the entire U.S. range for species that occur outside the eastern part of the country. The range maps are based on historical records, although ranges have retracted for many species in recent years as snake populations have declined. The shaded range should be viewed as an approximation of the actual presence of a species, as almost no snake has a continuous distribution across all habitats within a region. As an example, the range map for ring-necked snakes indicates that they occur throughout the eastern United States. Their actual distribution is patchy, however, as a result of their habitat requirements and the disappearance of populations due to natural or human-based causes.

HOW TO USE THE SPECIES ACCOUNTS

The species entries are arranged as follows (not all elements occur with every entry):

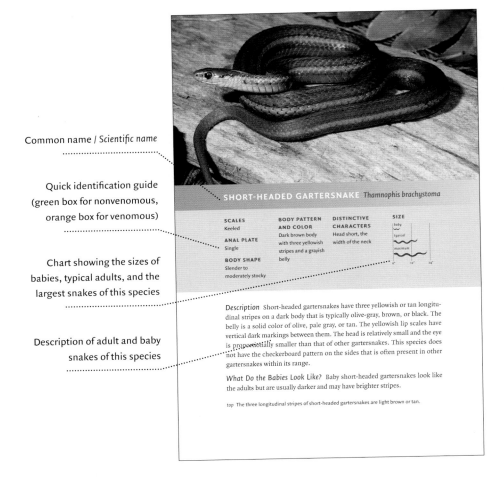

Common name / Scientific name

Quick identification guide (green box for nonvenomous, orange box for venomous)

Chart showing the sizes of babies, typical adults, and the largest snakes of this species

Description of adult and baby snakes of this species

SHORT-HEADED GARTERSNAKE *Thamnophis brachystoma*

SCALES	BODY PATTERN AND COLOR	DISTINCTIVE CHARACTERS	SIZE
Keeled	Dark brown body with three yellowish stripes and a grayish belly	Head short, the width of the neck	baby / typical / maximum
ANAL PLATE Single			
BODY SHAPE Slender to moderately stocky			

Description Short-headed gartersnakes have three yellowish or tan longitudinal stripes on a dark body that is typically olive-gray, brown, or black. The belly is a solid color of olive, pale gray, or tan. The yellowish lip scales have vertical dark markings between them. The head is relatively small and the eye is proportionally smaller than that of other gartersnakes. This species does not have the checkerboard pattern on the sides that is often present in other gartersnakes within its range.

What Do the Babies Look Like? Baby short-headed gartersnakes look like the adults but are usually darker and may have brighter stripes.

top The three longitudinal stripes of short-headed gartersnakes are light brown or tan.

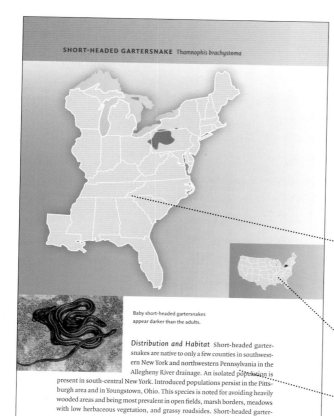

Map showing where this species is found in the eastern United States, with different colors for different subspecies and intergrades (see glossary). In the key, intergrades are indicated by an x between the subspecies.

Small map with a shaded area showing where this species is found in the continental United States

Where this species of snake lives

How to Use the Species Accounts 49

How this species behaves
and when it is active

How and what this species eats

How this species reproduces

What eats this species, and how
the snake protects itself

Is this species threatened
or endangered?

Information about
scientific classification

Interesting or anomalous
facts about snakes

snakes are often found beneath ground cover, including rocks, logs, or even discarded building materials.

Behavior and Activity Short-headed gartersnakes are active during the daytime in warm weather and hibernate underground during the winter. They have been observed basking in the open, but much of their time is spent under cover.

Food and Feeding Only earthworms have been documented as food for this species, although other small invertebrates are also potential prey.

Reproduction Short-headed gartersnakes mate in April and May, and several males may attempt to court a single female at the same time. Carl Ernst reported observing a "mating ball" of several males and females in Pennsylvania beneath a large rock in May. Litters average about 7 young (range, 4–14) and are born primarily in August.

Predators and Defense These little snakes presumably have a variety of predators, but none has been documented. Like many other snakes, they usually smear their captor with musk when picked up but rarely, if ever, even attempt to bite.

Conservation One of the greatest potential threats to this species is loss of habitat in its restricted geographic range. Destruction of a relatively small area could eliminate a large proportion of the existing populations. Within their geographic range, short-headed gartersnakes can be locally abundant, but some areas show signs of decline, such as along roads where grassy open areas suitable for basking have become overgrown with woody vegetation creating extensive shade.

What's in a Name? Edward Drinker Cope described this species in 1892 from a single specimen from Venango County, Pennsylvania, near the Allegheny River and named it *Eutaenia brachystoma*. The species name is derived from the Greek words *brachys* and *stoma*, which mean respectively "short" and "mouth," because this snake has fewer lip scales than previously described gartersnakes have. A 1908 treatise by Alexander Ruthven included this species within *T. butleri*, but in 1945 Albert G. Smith redescribed the short-headed gartersnake as a distinct species and named it *Thamnophis brachystoma*. The genus name is derived from *thamnos* and *ophis*, which mean respectively "shrub" and "snake" in Greek, referring to the frequent association of gartersnakes and ribbonsnakes with shrubby habitats.

78 *Small Terrestrial Snakes*

> **DID YOU KNOW?**
>
> *Four native species of eastern snakes (eastern and western wormsnakes, mud snake, and rainbow snake) have a pointed spine on the tip of their tails.*

small terrestrial snakes

previous page Red-bellied snake

SMOOTH EARTHSNAKE *Virginia valeriae*

SCALES
Smooth

ANAL PLATE
Usually divided

BODY SHAPE
Stout

BODY PATTERN AND COLOR
Solid brown or gray above with whitish belly

SIZE

Description The smooth earthsnake is a small brown to brownish gray snake with a light-colored belly. Its head is small and the nose is somewhat pointed. The scales are smooth, or some may be very weakly keeled, and a scattering of tiny dark spots is often visible on the back. Sometimes these dots are arranged into faint stripes. Three subspecies of the smooth earthsnake have been described (eastern smooth earthsnake, *V. v. valeriae*; western smooth earthsnake, *V. v. elegans*; and mountain smooth earthsnake, *V. v. pulchra*), but the differences among them are rather obscure.

top Smooth earthsnakes have no keeled scales.

What Do the Babies Look Like? Baby smooth earthsnakes look like the adults but are often darker in color.

Distribution and Habitat Smooth earthsnakes are found in parts of all the eastern states south of New England, New York, Michigan, and Wisconsin. They inhabit pine or hardwood forests, including the margins of swamps and other wetlands. They can be found in fields and rural areas adjacent to woodlands and in mountainous areas in the Appalachians. They are sometimes found in suburban areas.

Behavior and Activity Unless temperatures are extremely cold, smooth earthsnakes can be active year-round, including during warm spells in winter, but they usually remain beneath ground litter rather than on the surface. They are active aboveground primarily in early evening or at night during hot times of the year, and are notably active after rainstorms in Ohio and possibly in other areas.

left Tiny dark spots are visible on the back of some smooth earthsnakes.

below Smooth earthsnakes have an unmarked white belly.

SMOOTH EARTHSNAKE *Virginia valeriae*

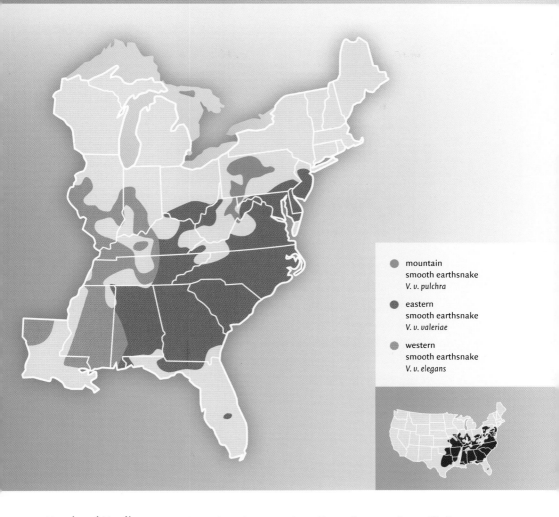

- mountain smooth earthsnake *V. v. pulchra*
- eastern smooth earthsnake *V. v. valeriae*
- western smooth earthsnake *V. v. elegans*

Food and Feeding Smooth earthsnakes eat primarily earthworms but will also eat the adults, larvae, and eggs of small, soft-bodied insects as well as slugs and small snails.

Reproduction Smooth earthsnakes mate in the spring and early summer, but nothing is known about their courtship. Presumably, mates find one another by following pheromone trails. They are livebearers that give birth to 2–14 (usually about 6) young between midsummer and early fall.

Smooth earthsnakes have brown heads and white lips.

Predators and Defense Because of their small size, smooth earthsnakes are probably eaten by a variety of reptiles, birds, and mammals. Known snake predators are kingsnakes, coral snakes, and racers. Smooth earthsnakes seldom if ever bite humans when picked up, although they may release musk from scent glands in the cloaca (anal region). Their defensive tactics include playing dead, curling the lips to display the teeth, and contorting the body to avoid being swallowed by a larger snake.

Conservation No specific conservation threats are associated with smooth earthsnakes, although they can be common in suburban areas and many are killed by domestic cats.

What's in a Name? This species was described by S. F. Baird and C. Girard in 1853 in the *Catalogue of North American Reptiles in the Museum of the Smithsonian Institution*. The name *valeriae* honors Valeria Blaney, who provided the specimen that the description was based on from Kent County, MD. Other specimens that were mentioned at the time were from Washington, DC, and Anderson, SC. None of the specimens was from Virginia, which the authors enigmatically chose as the genus name. The mountain smooth earthsnake, *V. v. pulchra*, has been proposed as a separate species but is not recognized as distinct by most herpetologists.

ROUGH EARTHSNAKE *Haldea striatula*

SCALES
Keeled

ANAL PLATE
Usually divided

BODY SHAPE
Moderately stout

BODY PATTERN AND COLOR
Solid brown above with whitish belly

SIZE
baby
typical
maximum
0"   12"   24"

Description Rough earthsnakes are small brown to grayish snakes with a light-colored belly, a small head, and a pointed nose. The scales are keeled, and there is often an indistinct light brown or tan band visible across the back of the head. The neck band and keeled scales distinguish the rough earthsnake from the smooth earthsnake (*V. valeriae*), which is superficially similar in appearance.

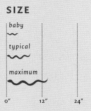

top A rough earthsnake from the Okefenokee Swamp in Georgia

left Some rough earthsnakes have a light band on the back of the head.

ROUGH EARTHSNAKE *Haldea striatula*

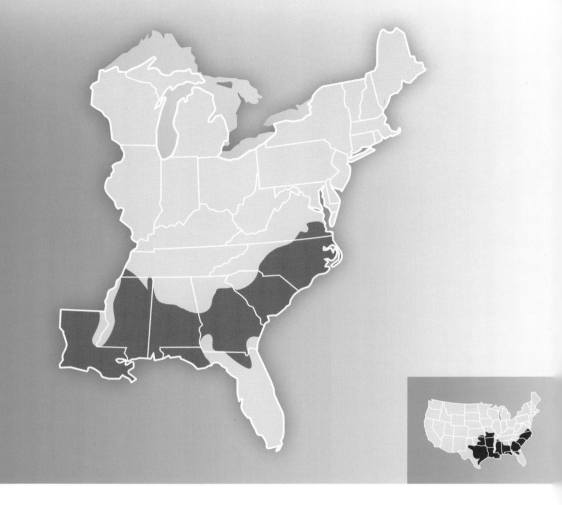

What Do the Babies Look Like? Baby rough earthsnakes look like the adults but are usually slightly darker and sometimes nearly black.

Distribution and Habitat Rough earthsnakes are found in a wide variety of habitats, including pine and hardwood forests, in parts of all eastern coastal states south of Virginia as well as in Tennessee. They are sometimes found in urban and suburban areas, including backyards and local parks.

Behavior and Activity Rough earthsnakes hibernate during cold winters but may become active during warm spells. They occasionally venture aboveground

Rough earthsnakes have keeled body scales.

in the daytime during spring but are more likely to appear around dusk or at night during summer, especially during rainy spells.

Food and Feeding Rough earthsnakes readily prey on earthworms, which they grab and swallow alive. They are known to eat beetle larvae and may also eat the eggs of ants and termites.

Reproduction Rough earthsnakes mate in the spring and early summer, but nothing is known about their courtship and mating behavior. Presumably, mates find one another by following pheromone trails. Earthsnakes give birth to an average of about 9 young (2–14) from midsummer into October.

Predators and Defense Several species of larger snakes prey on rough earthsnakes, including kingsnakes, coral snakes, and racers. Because outdoor domestic cats are known predators, they are likely also prey for common predators such as raccoons, opossums, and skunks. They seldom if ever bite a person if picked up but will sometimes release musk from scent glands in the cloaca (anal region).

Conservation No specific conservation threats are associated with rough earthsnakes, although domestic cats reportedly kill large numbers in suburban areas.

> **DID YOU KNOW?**
>
> Florida has 45 species of native snakes, more than any other eastern state. Maine has the fewest (10), of which one (timber rattlesnake) is now believed to be extinct in the state.

What's in a Name? Carl Linnaeus named this snake *Coluber striatulus* in 1766 based on a specimen he referred to as being from "Carolina." The species name is derived from Latin *striato*, meaning "striped," and *ula*, meaning "little" (presumably referring to the small band across the neck in the original specimen). The species was retained in the genus *Coluber* until 1820, when the German naturalist Blasius Merrem referred it to the genus *Natrix*, which comprised the watersnakes. Merrem was one of the first scientists to recognize reptiles and amphibians as being biologically distinct from each other. In 1853 S. F. Baird and C. Girard placed the rough earthsnake in the genus *Haldea*, named in honor of Samuel Stehman Haldeman, a nineteenth-century naturalist at the University of Pennsylvania. Over the next 150 years various authorities put the rough earthsnake in six other genera, including the genus *Virginia*, which includes the smooth earthsnake. Some authorities continue to place this species in the genus *Virginia*.

A rough earthsnake from Florida. This species varies little in appearance throughout its range.

BROWNSNAKE *Storeria dekayi*

SCALES
Keeled

ANAL PLATE
Divided

BODY SHAPE
Moderately stout

BODY PATTERN AND COLOR
Two rows of dark spots on gray or brown body

SIZE
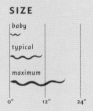

Description Brownsnakes are brown, grayish brown, or reddish brown above with two rows of small dark spots typically running the length of the back. They usually have small, obscure dark spots on the sides as well. Some individuals have a light stripe down the back. The belly is lighter with darker spots along the edges. Most individuals have a pair of larger spots on the neck that sometimes form a complete collar. Five subspecies occur in the eastern states (northern

top Some brownsnakes have a light stripe down the back.

left Brownsnakes usually have darker markings on the head and neck.

A midland brownsnake from near Terre Haute, Indiana

Newborn brownsnakes are tiny.

brownsnake, S. d. dekayi; marsh brownsnake, S. d. limnetes; Texas brownsnake, S. d. texana; midland brownsnake, S. d. wrightorum; and Florida brownsnake, S. d. victa). The midland subspecies intergrades with the northern subspecies over a broad geographic area and with the Texas subspecies over a narrow band (see map). The Florida brownsnake has a broad, light collar on the back of the head.

What Do the Babies Look Like? Baby brownsnakes are similar to the adults but the body color is usually darker (sometimes gray) with a pale yellow, grayish, or cream-colored neck ring.

Distribution and Habitat Brownsnakes are found in every eastern state and in many types of woodland habitats, including hardwood and pine forests, elevated or seasonally dry areas in swamps, and the margins of wetlands. They are common in wet floodplains and along creek bottoms. Brownsnakes adapt well in many suburban areas and can be abundant in flowerbeds, grass clumps, and vacant lots. In some regions, brownsnakes are actually encountered more frequently in residential areas and city parks than in natural forests.

Behavior and Activity Brownsnakes hibernate during extreme cold spells in the winter but may be active anytime from early spring to late fall, and even during winter warm spells. They sometimes crawl aboveground as well as beneath leaf litter, pine straw, and grass or other surface vegetation. They may be active during the day but are especially active at night and during rainy periods. They sometimes climb into low vegetation, including small trees.

Food and Feeding Brownsnakes eat mostly earthworms and slugs but will also eat snails, various insects, small salamanders, and spiders. They simply grab and swallow their prey alive. When feeding on snails, they pull the snail from its shell before swallowing it.

BROWNSNAKE *Storeria dekayi*

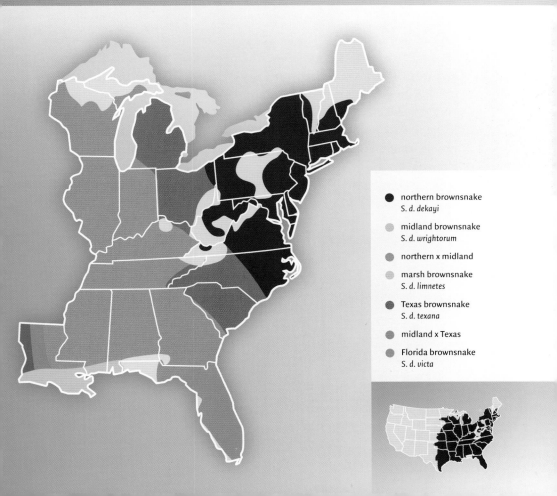

Reproduction Brownsnakes mate in the spring (March–May) and give birth to 3–41 young (average, about 13) during the summer or early fall. Farther south, mating and birth may occur earlier. In contrast to most reptile embryos, which are enclosed within a membrane and receive nourishment from yolk as they develop inside the mother, embryonic brownsnakes may receive nourishment from a placenta-like structure similar to that of mammals.

Predators and Defense Documented natural predators of brownsnakes include mammals (e.g., shrews, raccoons, opossums, and skunks), birds (e.g., robins, shrikes, brown thrashers, and hawks), snakes (e.g., racers, kingsnakes, and gartersnakes), and even toads and spiders. Brownsnakes respond to threats in a variety of ways. They first attempt to escape; if that fails, they may play dead, hide the head beneath the body coils, or flatten the body to appear larger. Brownsnakes are typically mild mannered when handled and rarely bite, although they may release a foul musk from their anal glands.

Conservation Many herpetologists consider the Florida brownsnake to be threatened in some areas due to extensive development within its restricted geographic range.

What's in a Name? Despite its widespread occurrence, being known from every eastern state, this species was not recognized scientifically until John Edwards Holbrook described it in 1839 as *Coluber dekayi*. He named the new species after Dr. James E. DeKay, a noted physician and naturalist from New York. Many herpetologists still refer to this species by its original common name, DeKay's snake or DeKay's brownsnake. The name sometimes causes problems in discussions with non-herpetologists who mistakenly hear the name as "decayed" snake. Some herpetologists consider the Florida brownsnake to be a separate species (*S. victa*) rather than a subspecies of *S. dekayi*.

Some brownsnakes, like this one from Ohio, are reddish in color.

RED-BELLIED SNAKE *Storeria occipitomaculata*

SCALES
Keeled

ANAL PLATE
Divided

BODY SHAPE
Slender to moderately stout

BODY PATTERN AND COLOR
Solid (sometimes with a stripe down the back) brown, black, gray, or orange above; belly red or orange

SIZE
baby
typical
maximum
0" 12" 24"

Description Red-bellied snakes are extremely variable in appearance. They may be gray, reddish, or brown, with three light spots on the neck sometimes forming a light ring that contrasts with a darker head. The belly is typically solid red, reddish brown, or orange; rarely an individual will have a nearly solid black belly. A light stripe may or may not run the length of the back. Two subspecies occur in the eastern states

top The bright red belly of this northern red-bellied snake from Maine is characteristic of the species throughout its geographic range.

Red-bellied snakes sometimes curl their upper lip when threatened.

RED-BELLIED SNAKE *Storeria occipitomaculata*

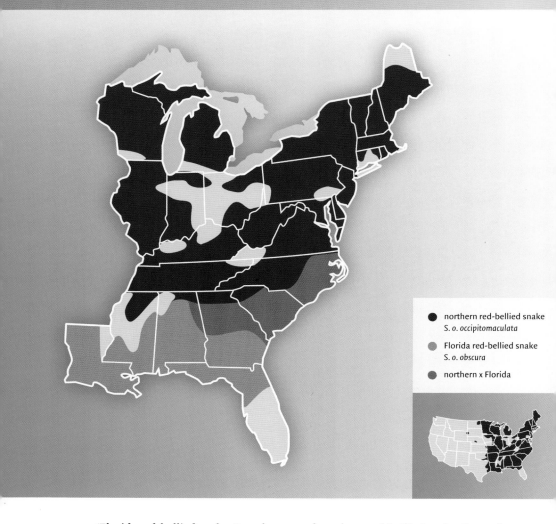

(Florida red-bellied snake, S. o. obscura; and northern red-bellied snake, S. o. occipitomaculata). The differences between the two subspecies are not very apparent, and a broad zone of intergradation occurs from Virginia to Alabama (see map).

What Do the Babies Look Like? Baby red-bellied snakes are similar to the adults but may be darker.

Distribution and Habitat Red-bellied snakes are found in portions of all eastern states in hardwood and pine forests, the margins of swamps and other wetlands, and marshy or boggy habitats. In mountainous areas they are often found in open habitats. They seem to be less common in developed areas than their close relatives, the brownsnakes.

Behavior and Activity Throughout most of their eastern range, red-bellied snakes are active during all warm months and during warm periods in winter. During extended cold periods they hibernate underground or beneath logs, rocks, and ground litter. Although they often remain hidden beneath leaves or other cover, they occasionally move about aboveground during early evening, and at night in the summer. They are noted for relatively long-distance travel, with some being recorded moving several hundred feet at a time—quite a distance for a small snake.

Food and Feeding Red-bellied snakes apparently eat mostly slugs and earthworms, which they swallow alive. They may also feed on soft-bodied insects, snails, and isopods (pillbugs or "roly-polies").

Reproduction Red-bellied snakes presumably mate in the spring but also have been observed mating during summer and fall. Nothing is known about

DID YOU KNOW?

Of the 63 species of snakes native to the eastern United States, only 7 are found in every one of the eastern states.

Red-bellied snakes vary greatly in body color at the same locality, and can be gray, reddish, or brown.

This red-bellied snake with reddish body color from Berkeley County, South Carolina, is an intergrade between the northern and Florida subspecies.

their courtship. The female typically gives birth to 4–9 (rarely, up to 23) live young from May to June in Florida, and in mid-to-late summer and fall in more northerly states.

Predators and Defense Known or suspected natural predators of these small snakes are those characteristic of woodland habitats and include snakes (e.g., kingsnakes and racers), birds, large spiders, toads, and salamanders. A peculiar response of many red-bellied snakes when they are picked up is to curl the upper lip to reveal a series of alternating black and white lines that sometimes look like large teeth and might appear threatening to a small predator. They may also respond with a combination of behaviors including flattening the body, releasing musk, and playing dead. Red-bellied snakes are generally inoffensive toward humans and seldom if ever bite when handled.

Conservation The destruction of their woodland habitats and small wetlands poses a major threat to red-bellied snakes. Domestic cats are also known to kill these small snakes in suburban communities.

What's in a Name? This species was called *Coluber occipito-maculata* by David Humphreys Storer, who described it in 1839 from a specimen from Amherst, Massachusetts. He called it the spotted-neck snake and selected the scientific name based on the Latin words *occiput* and *maculata* for "back of the head" and "spotted" respectively. In 1853 Baird and Girard placed this species along with *S. dekayi* in the genus *Storeria*, which was named in honor of Storer himself, who later became dean of the Harvard Medical School.

LINED SNAKE *Tropidoclonion lineatum*

SCALES
Keeled

ANAL PLATE
Single

BODY SHAPE
Moderately stout

BODY PATTERN AND COLOR
Solid gray, brownish, or olive above with three light-colored longitudinal stripes

DISTINCTIVE CHARACTERS
Light-colored belly with a pair of black semicircular spots on each belly scale

SIZE

Description The lined snake is a small gray, brownish, or olive snake with a distinctive yellow, white, or orange stripe down the center of the back and a light stripe down each side. The light-colored belly has a pair of dark semicircular spots in the center of each belly scale. The head is small and the eyes are comparatively smaller than those of gartersnakes and ribbonsnakes, which they superficially resemble. The scales are keeled.

What Do the Babies Look Like? Baby lined snakes have a more grayish body and the stripes are not as vivid as they are on adults.

Distribution and Habitat This is a snake of the central U.S. prairies. In the East, lined snakes are found only in Illinois and Wisconsin, in grasslands habitat that was originally prairie. In addition to small areas of remaining natural

top Lined snakes are small and moderately stout-bodied.

habitat, this species is most often associated with vacant lots or trash heaps in urban areas. Lined snakes are characteristically found in areas with rocks, boards, or other ground cover that they hide beneath.

Behavior and Activity Lined snakes become active in Illinois in March and remain active until cold weather in late October and November. They are secretive snakes that usually remain beneath ground litter but may be active on the surface at night during the summer and after periods of heavy rain.

Food and Feeding Lined snakes eat primarily earthworms. They have also been reported to eat sowbugs (roly-polies); small, soft-bodied insects; and insect larvae.

Reproduction In contrast to most other snakes, which mate in the spring, lined snakes mate in late August in Illinois, but nothing is known about their courtship behavior. They are livebearers that give birth to 2–17 young; typical litter sizes are 5-10. The babies are usually born in August or early September.

Predators and Defense Because of their small size, lined snakes probably fall prey to a variety of reptiles, birds, and mammals that happen to encounter them. Racers and loggerhead shrikes are among the predators that have been documented. These inoffensive snakes seldom if ever bite humans when picked up, although they may defecate and release musk from scent glands in the cloaca (anal region). When threatened they may flatten the body and curl the tail.

left Lined snakes give live birth to up to 17 babies.

below The two rows of semicircle-shaped spots running down a light belly is a diagnostic feature of lined snakes.

LINED SNAKE *Tropidoclonion lineatum*

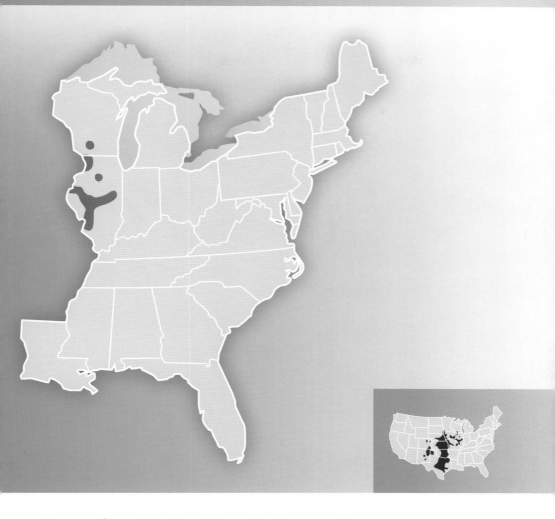

Conservation The lined snake has suffered from loss of its prairie habitat and is considered to be a threatened species at the state level in Illinois.

What's in a Name? Edward Hallowell described this species in 1856 based on a specimen from Kansas. The genus name is derived from the Greek words *tropis*, meaning "keel," and *klon*, meaning "twig," referring to the snake's keeled body scales and small size. The Latin word *lineatus*, meaning "of a line," refers to the obvious longitudinal stripes. Three subspecies of the lined snake have been described based on minor differences, but most herpetologists discount any subspecies designations. The herpetological community has also not accepted a suggestion that the species be included within the genus *Virginia*.

KIRTLAND'S SNAKE *Clonophis kirtlandii*

SCALES
Keeled

ANAL PLATE
Divided

BODY SHAPE
Moderately stout

BODY PATTERN AND COLOR
Gray to reddish brown back with four rows of rounded, dark blotches that may be indistinct; bright to pale red or pink belly with row of dark spots on sides

SIZE

Description The back of these small snakes ranges from grayish to reddish brown. Four parallel lines of dark, rounded spots, sometimes indistinct, run the length of the body. Spots on each side of the middle of the back alternate with those on the sides. A row of very distinct black spots runs along a cream-colored border on each side of the usually red or pink belly. The head is dark and the throat white or yellowish. Females get larger than males.

What Do the Babies Look Like? Baby Kirtland's snakes have a darker red belly and darker, more solid-colored back than adults.

top A Kirtland's snake from Clay County, Indiana, showing the pink belly and row of dark spots on the side

A newborn Kirtland's snake from Champaign County, Ohio

KIRTLAND'S SNAKE *Clonophis kirtlandii*

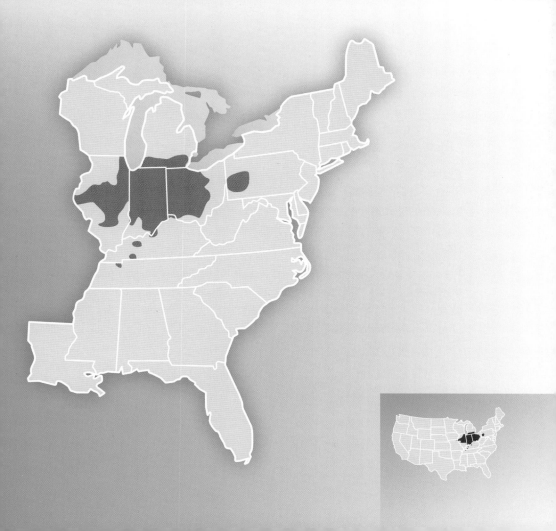

Distribution and Habitat The geographic range of Kirtland's snake is centered in most of Indiana and Ohio up to southern Michigan. The range extends across Illinois and into northwestern Kentucky, and in 2012 the species was discovered in a single county (Henry) in northern Tennessee. Isolated populations occur in Pennsylvania. The habitat includes swamp and marsh margins, woodlands, and open fields and meadows. Kirtland's snakes are characteristically found in moist areas near streams or isolated wetlands where crayfish burrows are often present. They are noted for inhabiting urban areas, including parks and vacant lots.

Behavior and Activity Kirtland's snakes are seen most commonly in the spring during April and May and are noticeably active again in autumn, especially October, but individuals can be found every month of the year. They are known to enter hibernation during November, but they may be seen during any month with moderate temperatures and have been observed basking on warm days during winter. Kirtland's snakes are usually found beneath rocks, logs, or other objects and are known to retreat into crayfish burrows in some localities. During summer they are usually active at night.

Kirtland's snake has a dark head and white or yellowish throat.

Food and Feeding Earthworms and slugs are common prey of Kirtland's snakes. Other invertebrates, including small crayfish, leeches, and a water strider (insect), have also been reported in their diet.

Reproduction Kirtland's snakes have been observed mating in April and May, but courtship has also been reported in early fall. They are livebearers that have 4–15 (average, about 8) young in late July to late September. As with many other snakes, larger females typically have larger litters.

Predators and Defense Kingsnakes and bullfrogs are known predators of Kirtland's snakes, but a variety of other species probably prey on these small snakes when given the opportunity. Among native wildlife proposed as potential predators are milksnakes, hawks, owls, and shrews. Larger carnivores such as raccoons, skunks, and foxes are also likely predators. Kirtland's snakes have a characteristic defensive posture in which they flatten the body to appear larger. They do this to a greater extent than most other snakes that exhibit similar behavior. Newborns will often coil, gape their mouth, and strike. Many Kirtland's snakes release an unpleasant-smelling musk when captured.

Kirtland's snakes are sometimes abundant in parks and vacant lots in urban areas.

Conservation Kirtland's snake populations appear to be declining throughout their geographic range. The species is listed as endangered (Indiana, Michigan, and Kentucky) or threatened (Illinois and Ohio) by most of the states where it occurs. Extensive urban development and agricultural activities that eliminate or degrade natural habitats are considered major contributors to the loss of populations. These snakes and the prey they depend on are also vulnerable to the introduction of pesticides into their environments. Kirtland's snakes persisting in urban areas likely represent relicts of larger populations that were once widespread in the region.

What's in a Name? Officially described as *Regina kirtlandii* in 1856 by Robert Kennicott, the species was named in honor of Jared Potter Kirtland, a nineteenth-century Ohio naturalist and physician for whom Kirtland's warbler is also named. The first of several known specimens of Kirtland's snake were from northern Illinois, where it was considered rare. In 1888 Edward Drinker Cope placed this species in the genus *Clonophis*. *Clono* is derived from the Greek *klon*, which means "twig" and presumably referred to the snake's small size. For most of the twentieth century this species was placed in the same genus (*Natrix*) as the watersnakes and referred to as Kirtland's watersnake.

SHORT-HEADED GARTERSNAKE *Thamnophis brachystoma*

SCALES
Keeled

ANAL PLATE
Single

BODY SHAPE
Slender to moderately stocky

BODY PATTERN AND COLOR
Dark brown body with three yellowish stripes and a grayish belly

DISTINCTIVE CHARACTERS
Head short, the width of the neck

SIZE
baby
typical
maximum
0"   12"   24"

Description Short-headed gartersnakes have three yellowish or tan longitudinal stripes on a dark body that is typically olive-gray, brown, or black. The belly is a solid color of olive, pale gray, or tan. The yellowish lip scales have vertical dark markings between them. The head is relatively small and the eye is proportionally smaller than that of other gartersnakes. This species does not have the checkerboard pattern on the sides that is often present in other gartersnakes within its range.

What Do the Babies Look Like? Baby short-headed gartersnakes look like the adults but are usually darker and may have brighter stripes.

top The three longitudinal stripes of short-headed gartersnakes are light brown or tan.

SHORT-HEADED GARTERSNAKE *Thamnophis brachystoma*

Baby short-headed gartersnakes appear darker than the adults.

Distribution and Habitat Short-headed gartersnakes are native to only a few counties in southwestern New York and northwestern Pennsylvania in the Allegheny River drainage. An isolated population is present in south-central New York. Introduced populations persist in the Pittsburgh area and in Youngstown, Ohio. This species is noted for avoiding heavily wooded areas and being most prevalent in open fields, marsh borders, meadows with low herbaceous vegetation, and grassy roadsides. Short-headed gartersnakes are often found beneath ground cover, including rocks, logs, or even discarded building materials.

Behavior and Activity Short-headed gartersnakes are active during the daytime in warm weather and hibernate underground during the winter. They have been observed basking in the open, but much of their time is spent under cover.

> **DID YOU KNOW?**
> *Four native species of eastern snakes (eastern and western wormsnakes, mud snake, and rainbow snake) have a pointed spine on the tip of their tails.*

Food and Feeding Only earthworms have been documented as food for this species, although other small invertebrates are also potential prey.

Reproduction Short-headed gartersnakes mate in April and May, and several males may attempt to court a single female at the same time. Carl Ernst reported observing a "mating ball" of several males and females in Pennsylvania beneath a large rock in May. Litters average about 7 young (range, 4–14) and are born primarily in August.

Predators and Defense These little snakes presumably have a variety of predators, but none has been documented. Like many other snakes, they usually smear their captor with musk when picked up but rarely, if ever, even attempt to bite.

Conservation One of the greatest potential threats to this species is loss of habitat in its restricted geographic range. Destruction of a relatively small area could eliminate a large proportion of the existing populations. Within their geographic range, short-headed gartersnakes can be locally abundant, but some areas show signs of decline, such as along roads where grassy open areas suitable for basking have become overgrown with woody vegetation creating extensive shade.

What's in a Name? Edward Drinker Cope described this species in 1892 from a single specimen from Venango County, Pennsylvania, near the Allegheny River and named it *Eutaenia brachystoma*. The species name is derived from the Greek words *brachys* and *stoma*, which mean respectively "short" and "mouth," because this snake has fewer lip scales than previously described gartersnakes have. A 1908 treatise by Alexander Ruthven included this species within *T. butleri*, but in 1945 Albert G. Smith redescribed the short-headed gartersnake as a distinct species and named it *Thamnophis brachystoma*. The genus name is derived from *thamnos* and *ophis*, which mean respectively "shrub" and "snake" in Greek, referring to the frequent association of gartersnakes and ribbonsnakes with shrubby habitats.

EASTERN WORMSNAKE *Carphophis amoenus*

SCALES
Smooth

ANAL PLATE
Divided

BODY SHAPE
Slender;
very small head

BODY PATTERN AND COLOR
Solid light brown or dark gray above; pinkish belly

DISTINCTIVE CHARACTERS
Spine on tip of tail; tiny eyes

SIZE
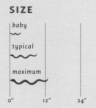

Description Eastern wormsnakes are either brown, dark gray, or nearly black above. The belly is either white or pink, and the light belly coloration extends slightly up onto the sides. The head is tiny and pointed, and there is a sharp point or spine on the tip of the tail. The eyes are extremely small and appear black. The two subspecies (eastern wormsnake, *C. a. amoenus*; and midwestern wormsnake, *C. a. helenae*) are difficult to tell apart without counting the head scales.

top The light pink or white belly coloration of the eastern wormsnake extends partway onto the sides.

left The eastern wormsnake has a pointed head and tiny black eyes.

EASTERN WORMSNAKE *Carphophis amoenus*

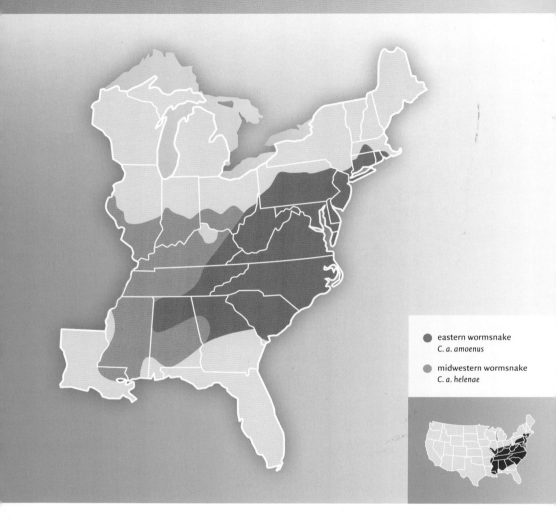

What Do the Babies Look Like? Eastern wormsnake babies resemble adults but are slightly darker.

Distribution and Habitat Eastern wormsnakes are found in portions of all eastern states south of Wisconsin, Michigan, New Hampshire, and Vermont, except Florida. Their habitat is typically cool, moist hardwood forests. They can be extremely abundant in the Piedmont of the Southeast, especially along rocky, wooded hillsides.

Behavior and Activity Eastern wormsnakes are active from early spring to late fall and in most areas hibernate in winter. They remain underground or beneath rocks and logs for most of their lives. Their occasional aboveground forays take place primarily at night during the warmest months.

Food and Feeding Eastern wormsnakes feed almost exclusively on earthworms, although they may rarely take soft-bodied insect larvae or slugs. When feeding, a wormsnake grasps an earthworm anywhere along its body, moves its mouth to one end or the other, and begins swallowing.

Reproduction Some evidence suggests that eastern wormsnakes may breed in the spring and again in the fall. They lay 1–12 eggs (usually 4 or 5) under rocks or in rotting logs in June or July, and the babies hatch in late summer or early fall. The eggs are often visible through the gravid (egg-carrying) female's somewhat translucent belly.

Predators and Defense Because they are among the smallest of the eastern snakes, wormsnakes are vulnerable to a variety of larger woodland snakes, birds, and mammals. When captured, wormsnakes wriggle vigorously and are unique among the small eastern snakes in pressing the spine of the tail against the captor, but never with enough force to penetrate the skin. They release a strong-smelling fluid from cloacal glands (anal region) when disturbed, but almost never bite humans.

top The tail of the eastern wormsnake has a spine-like tip.

bottom Eastern wormsnake eating an earthworm on a road in Scott County, Indiana

Conservation The loss of woodland habitat due to development is a potential threat to eastern wormsnakes, and insecticides have been implicated in population declines. Dead wormsnakes have been reported after floods in low-lying habitats, presumably because of their tendency to remain underground. Res-

The color of the body of eastern wormsnakes can be brown, dark gray, or almost black with smooth scales.

Wormsnakes almost never try to bite a person when picked up.

ervoirs or other damming projects that flood wooded areas could thus be detrimental to wormsnake populations.

What's in a Name? In the original description of this species in 1825, Thomas Say called this little snake "a very pretty and perfectly harmless serpent" that "inhabits Pennsylvania." He named it *Coluber amoenus*. In 1854 A. M. C. Duméril and G. Bibron placed the species in the genus *Carphophis*, a name derived from the Greek words *karphos*, meaning "twig," and *ophis*, meaning "snake," possibly referring to its small size. The species name is derived from the Latin word *amoenus*, meaning "beautiful" or "pleasant." Robert Kennicott described the midwestern subspecies in 1859, naming it *helenae* in honor of his cousin Helen Tennison of Monticello, Mississippi, who provided him with specimens of reptiles and amphibians, including wormsnakes. Some authorities refer to the eastern wormsnake species as the common wormsnake.

WESTERN WORMSNAKE *Carphophis vermis*

SCALES
Smooth

ANAL PLATE
Divided

BODY SHAPE
Slender; very small head

BODY PATTERN AND COLOR
Solid black above; pinkish belly

DISTINCTIVE CHARACTERS
Spine on tip of tail; tiny eyes

SIZE

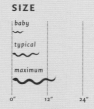

Description Western wormsnakes are black above with a pink belly. The belly coloration extends up onto the sides. The tiny head is distinctly pointed, the eyes are very small, and the tip of the tail ends in a sharp point or spine.

What Do the Babies Look Like? The babies resemble the adults.

Distribution and Habitat Western wormsnakes are found in parts of northern Louisiana, in extreme western Illinois, and in southwestern Wisconsin. The species is noted for occurring in cool, moist hardwood habitats, especially those associated with stream margins.

top Western wormsnakes are solid black above with a pink belly.

Behavior and Activity Western wormsnakes become active in early spring and remain so into late fall before hibernation. They seldom venture aboveground, remaining beneath rocks and other ground cover all year except for occasional nighttime forays during the summer, especially during rainy periods.

Food and Feeding The diet is restricted primarily to earthworms as well as occasional slugs and insect larvae. One western wormsnake was reported to have eaten a small ring-necked snake.

Reproduction Western wormsnakes breed in the spring and again in the fall and lay 1–12 eggs (usually 4 or 5) under rocks or in rotting logs in June or July. The eggs are often visible through the female's translucent belly. The young hatch in late summer or early fall.

Predators and Defense Because they are so small, western wormsnakes are prey for many snakes, birds, and mammals found in the same habitat. When captured, western wormsnakes wriggle vigorously in an attempt to escape and, like the closely related eastern wormsnake, may press their tail spine against the captor. They release a strong-smelling fluid from cloacal glands (anal region) when disturbed, but almost never bite humans.

Western wormsnakes are found east of the Mississippi River only in Illinois and Wisconsin.

WESTERN WORMSNAKE *Carphophis vermis*

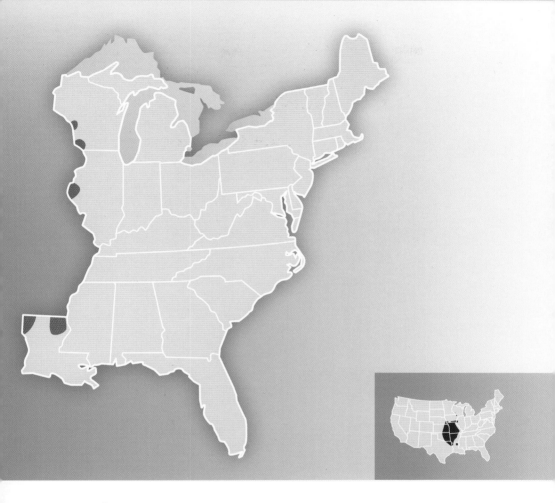

Conservation Clearing of woodland habitats for commercial development and agriculture can be detrimental to western wormsnake populations. Flooding of low-lying habitats after dam construction can also be detrimental because these little snakes tend to remain underground.

What's in a Name? Robert Kennicott described this species in 1859 based on a preserved specimen in the museum at Northwestern University that was "larger" and "much darker" than the two eastern wormsnakes, which at the time were themselves considered to be separate species. The Latin word *vermis*, meaning "worm," is descriptive of the body shape.

SOUTHEASTERN CROWNED SNAKE Tantilla coronata

SCALES
Smooth

ANAL PLATE
Divided

BODY SHAPE
Slender

BODY PATTERN AND COLOR
Tan or grayish brown; head and neck black with a light band just behind the head

SIZE
baby
typical
maximum
0" 12" 24"

Description This small, slender snake has a solid light brown or grayish brown back with a black head and neck that are separated by a light, cream-colored band. The belly is solid white or light pink. The head is rather pointed, and the lower jaw is countersunk.

What Do the Babies Look Like? Baby southeastern crowned snakes look like the adults.

Distribution and Habitat Southeastern crowned snakes are found in portions of all the southeastern states, Kentucky, and southern Indiana but are replaced in peninsular Florida by other species of crowned snakes. They are generally most

top The black head and neck of a southeastern crowned snake are separated by a light, cream-colored band.

SOUTHEASTERN CROWNED SNAKE *Tantilla coronata*

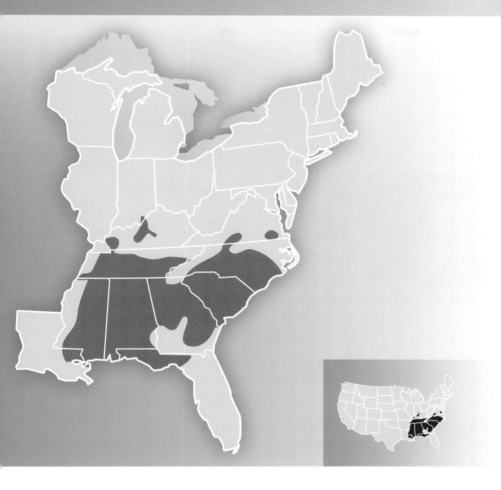

abundant in areas with sandy or other loose soil types and abundant surface litter, and may be found in both dry and moist woodland habitats.

Behavior and Activity Southeastern crowned snakes hibernate during winter but are likely to be active beneath ground litter on warm days in early spring and late fall. Most aboveground travel is during early evening or at night, but individuals are active beneath logs, rocks, and ground litter during the day in spring, summer, and fall. Crowned snakes are adept burrowers in sandy soil, almost appearing to swim into the sand when discovered.

Food and Feeding Southeastern crowned snakes eat a variety of prey, including centipedes, insect larvae, earthworms, and spiders. Small, grooved fangs at the

Southeastern crowned snakes eat centipedes.

A southeastern crowned snake from Telfair County, Georgia

back of the jaw direct venom into the prey. The weak venom's effectiveness in subduing prey is questionable, however, and all crowned snakes are presumably harmless to humans.

Reproduction Mating occurs in spring and late summer or early fall. Females that mate in the fall store the sperm until the following spring, when they produce one to three eggs in June or early July. The eggs hatch in the fall.

Predators and Defense Because of their small size, southeastern crowned snakes fall prey to most carnivorous vertebrates that occur in forested habitats, especially kingsnakes and coral snakes. They try to escape by burrowing into sand or soft soil, or by crawling beneath leaves or other ground litter. When captured, they do not bite but may release small quantities of musk from scent glands in the cloaca (anal region). Eastern bluebirds are known to prey on this species.

Conservation Except for degradation and destruction of their forest habitats, southeastern crowned snakes face no particular conservation threats.

What's in a Name? S. F. Baird and C. Girard described the species in 1853 based on a specimen from near Meridian in Kemper County, Mississippi. The name *Tantilla* is derived from the Latin word *tantillum*, meaning "a trifle." The species name is derived from the Latin word *coronata*, which means "crowned," referring to the black head and neck separated by a light band (crown).

RIM ROCK CROWNED SNAKE *Tantilla oolitica*

SCALES
Smooth

ANAL PLATE
Divided

BODY SHAPE
Slender

BODY PATTERN AND COLOR
Tan to reddish brown with a dark head

SIZE
baby
typical
maximum
0" 12" 24"

Description The rim rock crowned snake is slender and brown to reddish brown with a black head and a solid white to pinkish belly. The dark coloration on the head extends onto the neck, and sometimes an obscure, light-colored neckband is present.

What Do the Babies Look Like? Baby rim rock crowned snakes look like miniature adults.

Distribution and Habitat The rim rock crowned snake inhabits limestone outcrops and fossil coral reefs in the southeastern tip of peninsular Florida and the Florida Keys. It has the most limited geographic range of any eastern snake.

top The rim rock crowned snake inhabits limestone outcrops and fossil coral reefs.

Behavior and Activity Like other species of crowned snakes, this species is seldom seen aboveground but is presumed to be active all year in southern Florida beneath the rocky soil or ground litter. The rim rock crowned snake's diet has not been documented, but centipedes and other small invertebrates are presumed to constitute a major part of it.

Reproduction Scientists know virtually nothing about the reproductive biology of rim rock crowned snakes other than they are egg-layers. Their reproduction is presumed to be similar to that of other crowned snakes.

RIM ROCK CROWNED SNAKE *Tantilla oolitica*

Rim rock crowned snakes are found only in the southeastern tip of peninsular Florida and the Florida Keys.

Predators and Defense Rim rock crowned snakes are presumably prey for any predator larger than they are that can pursue and capture them underground. Although they have tiny rear fangs like other members of the genus *Tantilla*, these small snakes do not bite humans and are completely harmless.

Conservation Some herpetologists consider the rim rock crowned snake to be threatened because of the species' limited geographic range and the extensive urban development in southern Florida.

What's in a Name? S. R. Telford Jr. declared this species distinct from the southeastern crowned snake and Florida crowned snake in 1966 and named it for the oolitic limestone formations characteristic of its habitat. Some herpetologists dispute its validity as a separate species.

FLORIDA CROWNED SNAKE Tantilla relicta

SCALES
Smooth

ANAL PLATE
Divided

BODY SHAPE
Slender

BODY PATTERN AND COLOR
Tan to reddish brown with a dark head

SIZE
baby
typical
maximum

0"   12"   24"

Description The Florida crowned snake is slender and reddish brown with a black head, dark neck band, and solid white to pinkish belly. The three subspecies (peninsula crowned snake, *T. r. relicta*; coastal dunes crowned snake, *T. r. pamlica*; and central Florida crowned snake, *T. r. neilli*) are distinguished primarily by subtly different head and neck patterns. Knowing a specimen's locality is the surest way to determine its subspecies.

What Do the Babies Look Like? Baby Florida crowned snakes look like the adults.

Distribution and Habitat The Florida crowned snake is most commonly associated with sandy habitats such as pine and scrub oak sandhills, pine flatwoods,

top Florida crowned snakes, like this peninsula crowned snake subspecies, are generally associated with sandy habitats.

and hammocks in Florida. The species has also been reported from Lowndes County in southern Georgia.

Behavior and Activity Florida crowned snakes are presumed to be active during all but the coldest periods in winter but spend most of their time beneath the soil or under rocks and ground litter. They are seldom active aboveground.

Food and Feeding Florida crowned snakes are known to eat beetle larvae, and may also eat centipedes and snails. Like the other species of eastern crowned snakes they have enlarged rear teeth that presumably direct

A Florida crowned snake from Highlands County, Florida

Florida crowned snakes, like this coastal dunes crowned snake subspecies, have a black head and dark neck band.

above Florida crowned snakes have smooth scales.

left Florida crowned snakes are seldom seen aboveground.

venom into their prey. The rear fangs and toxic saliva may help to subdue prey, but these little snakes are harmless to humans.

Reproduction Scientists know little about reproduction in Florida crowned snakes. They lay eggs, and their reproduction is likely similar to that of other members of the genus Tantilla. Eggs are probably laid anytime from late spring until August.

Predators and Defense Like other species of crowned snakes, Florida crowned snakes are likely eaten by an array of predators that are able to find and capture small snakes underground or beneath ground litter, including other snakes such as coral snakes. They do not bite when handled.

Conservation Some populations of the Florida crowned snake are likely threatened by habitat loss.

What's in a Name? Until 1966, when S. R. Telford Jr. described it as a full species, the Florida crowned snake was considered a subspecies of the southeastern crowned snake (T. c. wagneri) that was found only in peninsular Florida. Telford based the distinction of this species from the southeastern crowned snake and of the three subspecies of Florida crowned snake from each other primarily

FLORIDA CROWNED SNAKE *Tantilla relicta*

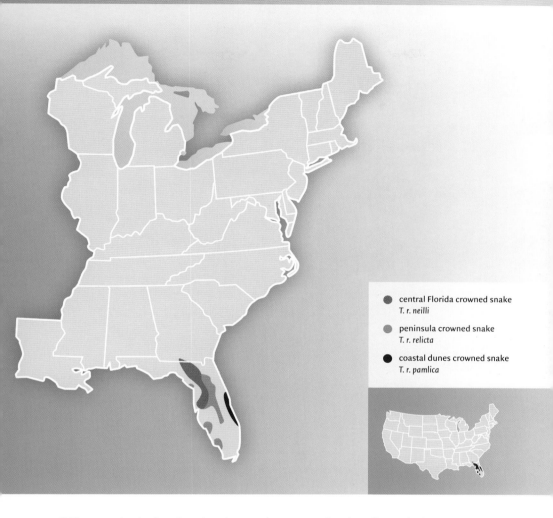

on differences in the hemipenis. The species name *relicta* is a direct derivative of the Latin word meaning "left behind" or "abandoned," which refers to the isolation of the Florida crowned snake from the wide-ranging southeastern crowned snake. The subspecies *T. r. pamlica* and *T. r. neilli* refer respectively to a geological formation in Florida called the Pamlico Terrace and to Wilfred T. Neill, known for his numerous contributions to the herpetology of Florida. All three subspecies of the Florida crowned snake are superficially similar in appearance to each other and to the southeastern crowned snake.

FLAT-HEADED SNAKE Tantilla gracilis

SCALES
Smooth

ANAL PLATE
Divided

BODY SHAPE
Slender

BODY PATTERN AND COLOR
Light brown or gray with dark gray or black head

SIZE

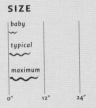

Description Flat-headed snakes are small, slender, brown or grayish snakes with a pink belly. As the name implies, the dark gray to black head is rather flattened. The head is also somewhat pointed, and the lower jaw is countersunk.

What Do the Babies Look Like? Baby flat-headed snakes look like the adults.

Distribution and Habitat This species occurs in only two eastern states: northwestern Louisiana and southeastern Illinois. It is usually found in hardwood or mixed hardwood-pine forests, especially in rocky areas, often under rocks and frequently associated with limestone habitats.

top Flat-headed snake

Behavior and Activity Flat-headed snakes are most active in spring and fall; they hibernate during winter and presumably remain underground during much of the summer. Aboveground activity usually occurs in early evening and at night during warm months.

Food and Feeding Like many members of the genus *Tantilla*, flat-headed snakes eat centipedes, beetle and other insect larvae, spiders, snails, and slugs. Slightly enlarged, grooved teeth at the back of the mouth direct weak venom into the

FLAT-HEADED SNAKE *Tantilla gracilis*

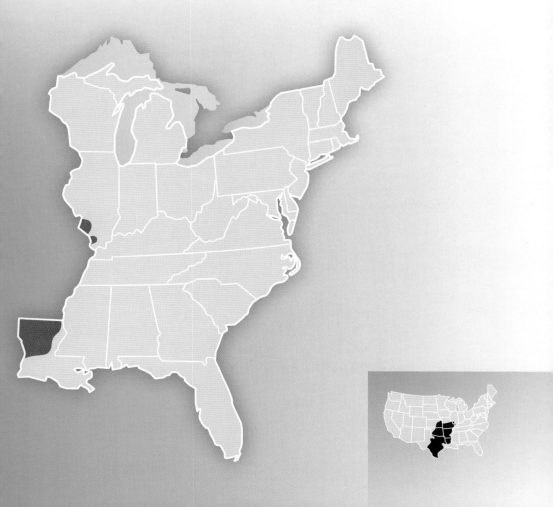

The flat-headed snake has a pinkish belly.

Baby flat-headed snakes are tiny.

prey while the snake is biting it. These snakes pose no danger to humans and rarely, if ever, bite when picked up.

Reproduction Mating occurs in spring, and up to four eggs are laid in June or early July, usually under rocks or in moist sand. The elongate, smooth-shelled eggs typically hatch in August or September.

Predators and Defense Native mammals, birds, and reptiles (snakes and lizards) are all natural predators of these little snakes. When disturbed, flat-headed snakes will try to escape by burrowing underground or beneath debris. If captured, they will not bite but will try to insert their snout between the captor's fingers and may release musk.

Conservation This species currently faces no particular conservation threats but is vulnerable to the same human-caused disturbances, including urbanization, that affect many animals.

What's in a Name? S. F. Baird and C. Girard described this species in 1853 based on a specimen U.S. Army colonel James Duncan Graham provided to the Smithsonian from Indianola on the Texas Gulf Coast. Indianola became a virtual ghost town after two powerful hurricanes made direct hits on it between 1875 and 1886. The species name is derived from the Latin word *gracilis*, meaning "slender."

SHORT-TAILED SNAKE *Lampropeltis extenuata*

SCALES
Smooth

ANAL PLATE
Single

BODY SHAPE
Very slender

BODY PATTERN AND COLOR
Primarily gray body with dark blotches on the back and sides sometimes interspersed with light orange or yellow blotches

DISTINCTIVE CHARACTERS
Very short tail

SIZE
baby
typical
maximum
0"   12"   24"

Description This very slender snake is gray with a row of dark brown to black blotches running along the center of the back and a row of smaller alternating dark brown or black blotches along each side. The gray background color is often interspersed with yellow, orange, or red, and the head is dark. The belly is gray to brown with small white spots.

What Do the Babies Look Like? Baby short-tailed snakes look like the adults.

Distribution and Habitat The short-tailed snake is known only from about 15 Florida counties along the Lake Wales Ridge that runs from the north-central region of the state near the Georgia border to the south-central region near Lake Okeechobee. Distribution records within the known geographic range are scarce because of this snake's secretive behavior and presumed rarity. The species is

top A short-tailed snake from Citrus County, Florida, with typical gray body and dark blotches on the back and sides.

confined almost exclusively to sandhill habitats of sand pine scrub, longleaf pine–turkey oak, and adjacent upland hardwoods, although a few specimens have been found in a sphagnum bog near sandhill habitat.

Behavior and Activity The short-tailed snake apparently spends most of its life underground and is seen so rarely that firm statements about most aspects of its behavior, including daytime activity periods, cannot be made. Most active aboveground individuals have been found during the day in spring (April and May) or fall (October and November).

Food and Feeding The natural diet is unknown. Captive specimens have eaten Florida crowned snakes, and presumably they prey on these and other small snakes, such as ring-necked snakes and brownsnakes, in the wild. They may also eat small lizards native to the region, including ground skinks, mole skinks, sand skinks, and worm lizards. The short-tailed snake is a constrictor.

Reproduction Short-tailed snakes lay eggs, but the clutch size is unknown, as are the details of courtship and mating, where the eggs are laid, and the incubation period.

Predators and Defense Coral snakes are known to prey on short-tailed snakes, and presumably other snake-eating snakes such as kingsnakes and racers are also natural predators. When captured, short-tailed snakes attempt to escape and may vibrate their tail.

Conservation Short-tailed snakes are so seldom seen that their actual status is difficult to ascertain. They appear to be very rare, but that may primarily reflect their extremely secretive habits. Conservationists consider their greatest threat to be loss of longleaf pine–turkey oak habitat in Florida due to urban, industrial,

Short-tailed snakes are extremely slender.

SHORT-TAILED SNAKE *Lampropeltis extenuata*

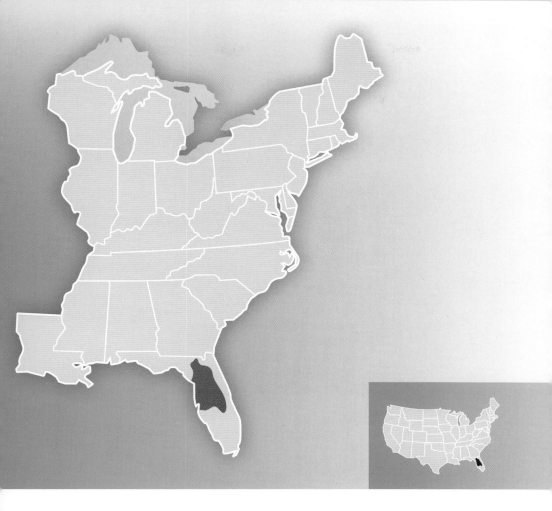

and agricultural development. Florida herpetologists would appreciate receiving news of any sightings, especially if documented with photographs.

What's in a Name? Arthur Erwin Brown, director of the Philadelphia Zoo, described this species as *Stilosoma extenuatum* based on a single specimen from the now extinct town of Lake Kerr in Marion County, Florida. He noted that the body was "very slender" and derived the original scientific name from words in both Greek (*stylos* and *soma*) and Latin (*extenuo*) that referred to the thin, columnar body. Some authorities continue to place this species in the genus *Stilosoma* rather than *Lampropeltis*. It is sometimes called the short-tailed kingsnake.

PINE WOODS SNAKE *Rhadinaea flavilata*

SCALES
Smooth

ANAL PLATE
Divided

BODY SHAPE
Slender

BODY PATTERN AND COLOR
Solid orange to reddish brown back with whitish belly

DISTINCTIVE CHARACTERS
Dark line passing through eye; lip scales white or yellowish

SIZE
baby
typical
maximum
0" 12" 24"

Description Pine woods snakes are light to reddish brown or dark orange above with a white to yellow unmarked belly. A faint narrow stripe may be present along the middle of the back. A dark brown stripe starts at the nose and extends through the eye to the corner of the mouth. The chin and lip scales are yellowish, white, or cream colored.

What Do the Babies Look Like? Baby pine woods snakes look like the adults.

Distribution and Habitat This species has a patchy geographic distribution in every coastal state from North Carolina to Louisiana, with large gaps apparently occurring between known populations. Although seldom common, pine woods snakes inhabit pine and mixed pine-hardwood forests. Individuals have

top Pine woods snakes are not restricted to living in pine forests.

been reported from hardwood hammocks in Florida, from pine areas on coastal islands, and even from inside houses.

Behavior and Activity Because they live mostly in warm coastal areas, pine woods snakes are active for much of the year, although they may hibernate underground or in logs during cold winters. They are most frequently encountered in the spring, but they are seldom seen aboveground and their daily activity patterns are unknown.

> **DID YOU KNOW?**
> Our perceptions of activity and rarity are inaccurate for some snake species because many spend more time underground than aboveground.

Food and Feeding Very little is known of the natural diet. Captive pine woods snakes prefer small frogs, salamanders, and an occasional small lizard. Enlarged teeth in the back of the mouth direct toxic saliva into the prey. The bite subdues their small prey animals but is of no consequence to humans.

Reproduction Little information exists regarding reproduction; mating apparently occurs in spring, and one to four (most often two or three) eggs are laid during summer. Some females may lay two clutches of eggs during a single year. The babies hatch 6–8 weeks later.

Predators and Defense Terrestrial snakes such as racers and kingsnakes have been reported to eat pine woods snakes, but other carnivorous pine forest animals such as shrews, birds, and toads are also potential predators. Pine woods snakes do not bite when picked up by humans but may release a bad-smelling musk.

The lip scales of a pine woods snake are white or yellowish.

PINE WOODS SNAKE *Rhadinaea flavilata*

Conservation Threats to this species are difficult to confirm because its secretive nature and scattered distribution make declines or disappearances extremely difficult to document. Habitat destruction, degradation, and fragmentation due to human activities are likely the greatest threats these snakes face.

What's in a Name? Edward Drinker Cope described this species in 1871 based on a specimen collected by Dr. Henry C. Yarrow near Fort Macon on the Outer Banks of North Carolina. Dr. Yarrow was an army surgeon from Philadelphia who was well known as an ornithologist and herpetologist. Cope named the new species *Dromicus flavilatus*, but in 1967 Charles W. Myers placed it in the

above Pine woods snakes are extremely secretive and rarely found.

left A dark orange pine woods snake from Alachua County, Florida

genus *Rhadinaea*. The name is derived from the Greek word *rhadinos*, meaning "slender," referring to the body shape. Cope derived the species name from Latin *flavus*, meaning "golden," and *lata*, meaning "wide," presumably referring to his description of the broad central portion of the back as being "a rich golden brown." Pine woods snakes are also known as yellow-lipped snakes or pine woods littersnakes.

RING-NECKED SNAKE *Diadophis punctatus*

SCALES
Smooth

ANAL PLATE
Divided

BODY SHAPE
Slender

BODY PATTERN AND COLOR
Gray or black, usually with light ring around neck; belly yellow to red, sometimes with small black dots

SIZE
baby
typical
maximum
0" 12" 24"

Description These very slender snakes are usually gray to nearly black above with a yellow, orange, or red belly and a light ring completely encircling the neck. In some subspecies the belly usually has a row of dark spots down the center. Five of the dozen subspecies occur in the eastern United States. The northern ring-necked snake (D. p. *edwardsii*) has a complete neck ring and only a few or no large spots on the belly. The southern ring-necked snake (D. p. *punctatus*) has a neck band that is broken in the middle and a row of conspicuous black spots running along the middle of the belly. The Mississippi ring-necked snake (D. p. *stictogenys*) has a very narrow, sometimes broken neck band and irregular spots along the center of the belly. The prairie ring-necked snake (D. p. *arnyi*) has numerous haphazardly placed dark spots on the yellow belly. The Key ring-

top The northern ring-necked snake subspecies has a complete neck ring and few spots on the belly.

RING-NECKED SNAKE *Diadophis punctatus*

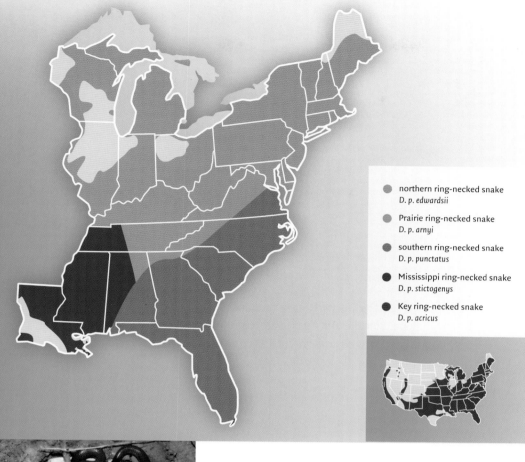

- northern ring-necked snake
  D. p. *edwardsii*
- Prairie ring-necked snake
  D. p. *arnyi*
- southern ring-necked snake
  D. p. *punctatus*
- Mississippi ring-necked snake
  D. p. *stictogenys*
- Key ring-necked snake
  D. p. *acricus*

Ring-necked snakes hatch from eggs in late summer or early fall.

necked snake (D. p. *acricus*), known only from the Florida Keys, has almost no neck ring.

What Do the Babies Look Like? Baby ring-necked snakes look like the adults.

Distribution and Habitat Ring-necked snakes are found in parts of all the eastern states, and throughout the entire state in most of them. They are commonly associated with wooded habitats such as mixed hardwood and pine

Ring-necked snakes are usually gray or nearly black above with a yellow, orange, or red belly.

forests in mountainous areas, swamp margins, and floodplains. The species is noted for requiring ground cover such as rocks, logs, or ground litter for hiding.

Behavior and Activity Ring-necked snakes are active throughout the year in southern Florida, and during all of the warm months elsewhere. They hibernate in areas with cold winters. These small snakes are noted for relatively long-distance travel—more than a mile in some instances and frequently more than 200 feet—but they seem more averse than many other snake species to crossing roads, particularly those with moderate to heavy traffic. Activity aboveground is primarily at night.

Ring-necked snakes eat a variety of small prey, including earthworms.

Food and Feeding The diet includes a variety of small animals such as earthworms, salamanders, small frogs, small snakes and lizards, and insect larvae. They may paralyze prey by chewing on it with their enlarged posterior teeth so that their toxic saliva can enter the lacerations.

Reproduction Mating occurs in spring and again in fall. Females may store the sperm from fall matings over the winter, and fertilization occurs in the spring. When courting, the male rubs his mouth along the neck of the female and bites her in the neck ring area. Ring-necked snakes lay two to seven eggs in moist, protected sites (e.g., in rotting logs and under rocks) during summer. Good nesting sites are sometimes used by more than one female. The eggs hatch in about 7–8 weeks.

Predators and Defense The number of known natural predators is high because of the extensive research that has been done on this abundant and widespread species. Eight species of terrestrial snakes have been reported to eat ring-necked snakes, as have five species of predatory birds, six native mammals, bullfrogs, and toads. The ring-necked snake's first response upon being captured is to attempt a rapid escape. When exposed while hiding beneath a log, the snake may momentarily flip over, exposing the bright yellow, orange, or red belly, and then turn back over and crawl into dark soil or rotting wood. Whether intentional or not, the maneuver can momentarily distract a predator (or a person) and allow the dark-bodied snake to escape. Ring-necked snakes also produce a musk that some people find extremely disagreeable. Some captured or disturbed individuals become motionless, as if dead; some may hide the head beneath the body; and some display the underside of the brightly colored tail, which may appear threatening to some small predators. These snakes seldom bite people and are generally considered harmless, but bites have sometimes been reported to cause a burning sensation.

Some ring-necked snakes display the underside of the brightly colored tail when threatened.

top The neck ring of Key ring-necked snakes is either very faint or completely missing.

left Some ring-necked snakes have a gray body.

Conservation Ring-necked snakes are extremely abundant in many areas, but like other species that require woodland habitats with a broad array of terrestrial prey, they are vulnerable to environmental alterations that destroy natural habitats or introduce pesticides into the ecosystem.

What's in a Name? This species, first recognized scientifically as *Coluber punctatus* by Carl Linnaeus in 1766 in the classic 12th edition of *Systema Naturae*, was based on a specimen from "Carolina." The species was later placed in the genus *Diadophis*, a name derived from Greek *diadema*, meaning "crown" or "headband," and *ophis*, meaning "snake," referring to the ring behind the snake's head. The species name is derived from the Latin word *punctum*, meaning "dot," and refers to the spotted belly.

previous page Checkered phase of the common gartersnake

SCARLET SNAKE *Cemophora coccinea*

SCALES
Smooth

ANAL PLATE
Single

BODY SHAPE
Slender

BODY PATTERN AND COLOR
Red, yellow or white, and black blotches or saddles on back; belly white

DISTINCTIVE CHARACTERS
Pointed red or orange nose

SIZE
baby
typical
maximum
0'   2'   4'

Description Scarlet snakes appear to have rings or bands, although the red, yellow or white, and black blotches that cover most of the back do not actually encircle the body. The pattern resembles that of the venomous coral snake, apparently a result of mimicry, except that the snout is not black like the coral snake's and the red bands do not touch the yellow bands. The belly is immaculate white, which readily distinguishes it from the scarlet kingsnake and coral snake. The pointed nose is always red or orange followed by a narrow black band. The two eastern subspecies (northern scarlet snake, *C. c. copei*; and Florida scarlet snake, *C. c. coccinea*) are distinguished by minor differences in scale numbers and spacing of the banding pattern.

top The red, pointed nose of a scarlet snake distinguishes it from a scarlet kingsnake or coral snake.

SCARLET SNAKE *Cemophora coccinea*

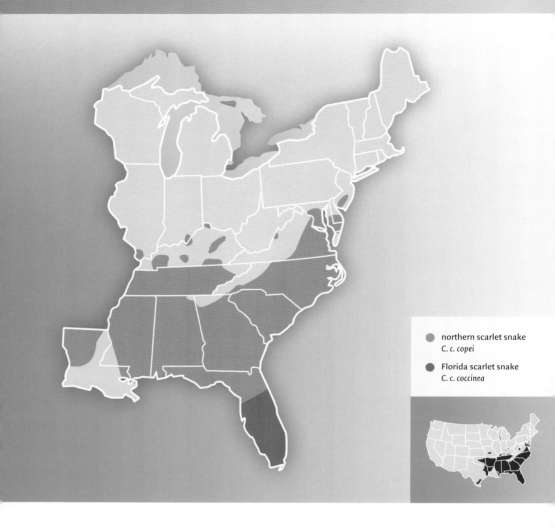

What Do the Babies Look Like? Baby scarlet snakes look like the adults.

Distribution and Habitat Scarlet snakes are found throughout much of the southern Atlantic and Gulf Coastal Plains and are more patchily distributed in the Piedmont of most states. The geographic range includes most areas outside of mountainous habitat from the District of Columbia, Maryland, and eastern Virginia to the Florida Keys, and across to Kentucky, southern Illinois, and southern Indiana. Isolated populations occur in the Pine Barrens of New Jersey. The species is apparently absent from the central and southern portions of Louisiana. Scarlet snakes are associated mostly with pine, hardwood, and mixed hardwood-pine forests, and they can be common in mangrove habitats in the Everglades. Generally, they prefer open forests with sandy soil and lots of ground litter and organic debris under which they can retreat. They are considered fossorial, spending most of their time underground, and use the pointed nose to burrow through loose soil.

Behavior and Activity Scarlet snakes are active aboveground almost exclusively at night and only during the warmer months. They are often found crossing roads at night or beneath logs, pine straw, and other ground cover.

Food and Feeding Scarlet snakes feed primarily on the eggs of other reptiles (snakes, lizards, and turtles), which they slit open with their enlarged, knife-like teeth. The snake feeds either by squeezing the egg to expel the contents or by sticking its head into the egg and drinking the liquid. Small reptile eggs may be eaten whole. They also use constriction to subdue and eat small lizards, snakes, and frogs.

Reproduction Very little is known about the reproductive biology of scarlet snakes. They are egg-layers that apparently mate in the spring and lay their elongated eggs during the summer in underground burrows, under rocks, in rotten logs, or in similar places. Clutches average about 5 eggs but can range from 2 to 13. The eggs hatch after about 2 months, usually during late summer or fall.

Predators and Defense The known natural predators are coral snakes, other snake-eating snakes, predatory birds, and mammals. Scarlet snakes seldom bite when captured although they sometimes release a mild-smelling musk.

top Scarlet snakes seldom bite when picked up. Note the solid white belly.

left Baby scarlet snakes are tiny and look like the adults.

Scarlet Snake 115

Scarlet snake from Franklin County, Florida

Conservation Two of the greatest threats to scarlet snakes result from fragmentation of their habitat by roads (they suffer extensive road mortality in areas of heavy nighttime traffic) and destruction of their natural habitats associated with commercial development.

What's in a Name? J. F. Blumenbach described the Florida scarlet snake as a distinct species in 1788, naming it *Coluber coccinea*. He based the name on the Latin word *coccinea*, which means "scarlet." The northern scarlet snake was recognized as a subspecies in 1863 and named *C. c. copei* in recognition of Edward Drinker Cope, who placed the species in the genus *Cemophora* in 1860. *Cemophora* is derived from two Greek words, *kemos* and *phoros*, respectively "muzzle" and "bearing," in reference to the pointed snout, which bears the force during the snake's excavations. Cope's most prestigious honor for his many contributions to herpetology and ichthyology came in 1913 when John T. Nichols of the American Museum of Natural History established a scientific journal "to advance the science of cold-blooded vertebrates" and named it *Copeia*.

ROUGH GREEN SNAKE *Opheodrys aestivus*

SCALES
Keeled

ANAL PLATE
Divided

BODY SHAPE
Extremely slender

BODY PATTERN AND COLOR
Solid green above; typically pale yellow or white below

DISTINCTIVE CHARACTERS
Body color turns blue after death

SIZE

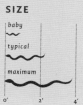

Description The rough green snake is very thin, almost vinelike in appearance and is solid green above. The belly is usually yellowish or white. The head is somewhat distinct, and the eyes are rather large. When a green snake dies, its skin usually turns bluish gray.

What Do the Babies Look Like? Baby green snakes are grayish green rather than bright green like the adults.

Distribution and Habitat Rough green snakes are found in parts of every eastern state south of Wisconsin, Michigan, and New York and throughout most of the southeastern states. They are expert climbers and are difficult to find in the bushes and vines they frequently inhabit. They can be especially common in trees and vines along the edges of water bodies and roads.

top A rough green snake from Marion County, Florida

Behavior and Activity Rough green snakes are active almost exclusively during the daytime hours throughout the warmer months and commonly spend their nights in trees, vines, or bushes, sometimes several feet above the ground. They typically hibernate underground or in stumps or logs during winter, and have even been found in ant mounds.

Food and Feeding The diet consists mostly of spiders, insects, and insect larvae, but also includes millipedes, land snails, and small treefrogs. Green snakes usually grab and swallow their prey alive but may chew on very active animals to subdue them before swallowing.

left The belly of a rough green snake is yellow or yellowish white.

below Rough green snakes eat small invertebrates, including butterflies.

ROUGH GREEN SNAKE *Opheodrys aestivus*

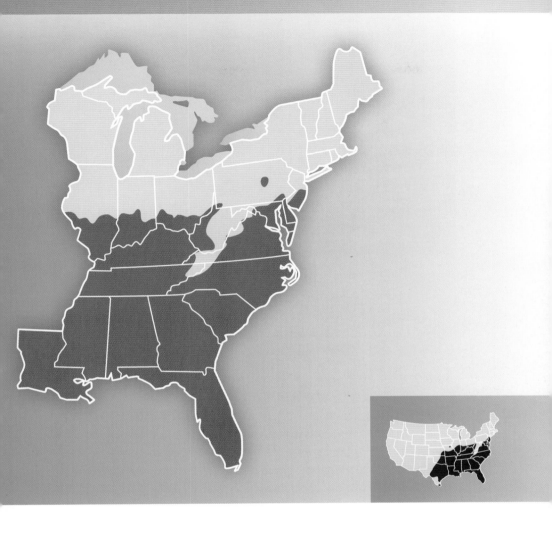

Reproduction Mating occurs primarily in the spring, but they may also mate in the fall. They commonly nest in hollow trees, laying 3–12 eggs, usually during the summer. Females may also lay eggs in rotting logs, under rocks, or in other concealed moist places. Several females may lay their eggs in the same location.

Predators and Defense Birds, spiders, and other snakes are documented predators. The rough green snake's bright green color is an ideal camouflage in vines, shrubs, and bushes, where the snake may remain completely motion-

Rough green snakes often climb high into trees and shrubs.

less and undetected while a person or predator is near. When captured, green snakes may open their mouth but seldom, if ever, actually bite.

Conservation Because rough green snakes are often associated with the margins of aquatic areas, development that destroys shrubbery and other vegetation around streams and wetlands may reduce available habitat for them. Likewise, certain nonnative ornamental plants may be less likely to attract insects, thereby reducing available prey. Because green snakes eat insects and spiders, they may be particularly susceptible to pesticides. Green snakes are often killed on highways adjacent to wetland or woodland habitats.

What's in a Name? Carl Linnaeus described this species in 1766 and named it *Coluber aestivus* based on a specimen from "Carolina." The word *aestivus* means "summer" in Latin; possibly Linnaeus viewed "Carolina" as a summery location. L. Fitzinger placed the genus in *Opheodrys* in 1843, a name derived from Greek *ophis*, meaning "snake," and *drymos*, referring to oak trees. Arnold B. Grobman of the University of Florida published a scientific paper in 1964 recognizing four subspecies of rough green snakes based on variations in scale measurements. However, the proposed subspecies show little variation in appearance from each other throughout their range, and most authorities do not consider the subspecific designations valid.

SMOOTH GREEN SNAKE *Opheodrys vernalis*

SCALES
Smooth

ANAL PLATE
Divided

BODY SHAPE
Slender to moderate

BODY PATTERN AND COLOR
Solid green above; white to yellowish white below

DISTINCTIVE CHARACTERS
Body bright green with smooth scales

SIZE
baby
typical
maximum
0"   12"   24"

Description The body is solid bright green above with a solid white or pale yellowish belly. Smooth green snakes are somewhat more robust than rough green snakes and have smooth scales. An uncommon tan color phase found in Wisconsin is light brown above.

What Do the Babies Look Like? Baby smooth green snakes are more grayish or olive than the bright green adults.

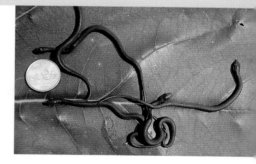

Baby smooth green snakes are not as brightly colored as the adults.

Distribution and Habitat The southernmost geographic range in the eastern states is limited to mountainous areas of northwestern Virginia, southern

top A smooth green snake from Greene County, Ohio

SMOOTH GREEN SNAKE *Opheodrys vernalis*

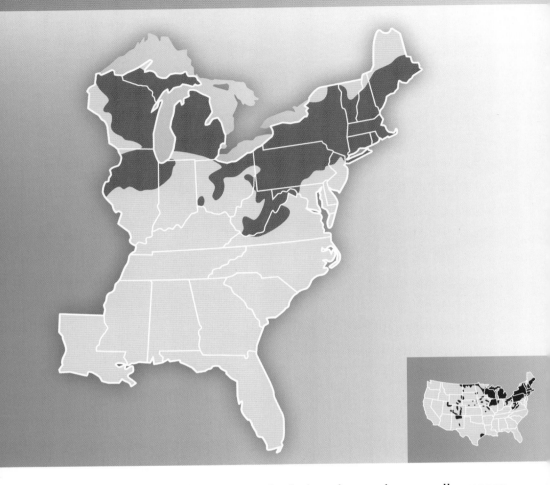

West Virginia, and western Maryland where these snakes generally occur on the ground in high-elevation moist meadows or grassy areas. They are found in Michigan and in parts of Wisconsin, Illinois, Indiana, Ohio, Pennsylvania, New York, and New Jersey but are in widely separated populations within each state. Smooth green snakes are found in all of the New England states but are absent from northern portions of Vermont, Maine, and New Hampshire. A single historical specimen thought to be from North Carolina has now been discredited. They are often found beneath rocks or other debris in grasslands, along road cuts through rocky terrain, and on the edges of ski slopes. They may enter the edges of mixed hardwood and conifer forests but are infrequently found in heavily wooded areas.

Behavior and Activity Smooth green snakes are active during the day throughout the warmer months. They do not typically climb as rough green snakes do. In winter they typically hibernate underground or in stumps or logs, and sometimes are found in ant mounds.

Food and Feeding Smooth green snakes eat insects and their larvae, spiders, centipedes, millipedes, worms, small salamanders, and small crayfish. They are not constrictors, but simply grab and swallow their prey.

Reproduction Smooth green snakes mate in the spring and possibly in the fall, and lay 2–13 eggs in rotting logs, under rocks, or in other concealed moist places, often in open areas with no tree canopy. The eggs go through extensive development while still inside the female and thus have a shorter incubation period than any other U.S. snake, hatching in late summer about 2 weeks or less after they are laid.

The nonkeeled (smooth) body scales readily distinguish a smooth green snake from a rough green snake.

Predators and Defense Snake-eating snakes such as kingsnakes, hawks, and spiders have been documented to eat smooth green snakes, which depend on their bright green color for camouflage in grassy areas. Smooth green snakes typically do not bite when picked up but will readily defecate.

Conservation Habitat destruction, including removal of large rocks, has been implicated as a major threat to this species and may eliminate some populations. Smooth green snakes are frequent victims of road mortality on highways near their habitats. Broad-scale applications of insecticides, including spraying to control gypsy moths in forested habitats where a variety of caterpillar species form a major portion of the diet, could be a major problem.

What's in a Name? Richard Harlan described this species and named it *Coluber vernalis* in 1826 on the basis of a preserved specimen at the Philadelphia Academy of Natural Sciences said to be found in Pennsylvania and New Jersey. The word *vernalis* means "of the spring" in Latin and was possibly chosen because the author associated what he described as the snake's "universal deep green colour" with springtime. Since its original description the smooth green snake has been assigned to eight different genera; most recently some herpetologists placed the smooth green snake as the only species in the genus *Liochlorophis* rather than in *Opheodrys* with the rough green snake.

PLAINS GARTERSNAKE *Thamnophis radix*

SCALES
Keeled

ANAL PLATE
Single

BODY SHAPE
Moderately stocky

BODY PATTERN AND COLOR
Body usually black, dark brown, or dark greenish but sometimes tan or brick red, and with three yellow stripes; belly pale yellow with row of dark spots along edges

DISTINCTIVE CHARACTERS
Center stripe bolder than side stripes; prominent dark vertical lines on yellow lip scales

SIZE
baby
typical
maximum
0'   2'   4'

Description Plains gartersnakes typically have three distinct yellow or orange stripes that run the length of the body. The body color is usually black or brown, sometimes with a greenish or olive hue, but some individuals have a tan or reddish body. The stripe down the center of the back is usually bolder and more orange than the side stripes, which are typically more yellowish. The side stripes usually include the third and fourth scale rows above the belly. Rows of dark spots run the length of the body between the center stripe and side stripes and between the side stripes and the belly, forming a sort of checkerboard pattern. Dark spots run along the outer edge of the belly scales. The belly is usually pale

top The black vertical lines on the lips of plains gartersnakes distinguish them from ribbon snakes, which have plain yellow lips.

PLAINS GARTERSNAKE *Thamnophis radix*

yellow. The yellow lip scales have distinct vertical dark markings. The tongue is typically red with a black tip.

What Do the Babies Look Like? Baby Plains gartersnakes look like the adults.

Distribution and Habitat Plains gartersnakes are found in northern Illinois and farther south in a population across the Mississippi River from Missouri; southern and west-central Wisconsin; northwestern Indiana; and an isolated area in central Ohio. Wet prairie habitats are preferred, especially near slow-moving water or ponds, the edges of drainage ditches, wet meadows, and pastures. They are occasionally found in wooded areas and around farms as well as in green spaces in urban areas. Plains gartersnakes are often found in association with crayfish burrows, which are known sites for hibernation and also serve as refugia in hot periods during summer.

Behavior and Activity Plains gartersnakes typically become active in the eastern states in March and remain aboveground into November. They are active primarily during warm periods of the day, but nighttime forays and feeding have been observed during the summer. They enter crayfish burrows, mammal burrows, or other underground retreats at night and for hibernation in late fall.

Food and Feeding Plains gartersnakes eat a variety of prey including frogs, salamanders, small fish, birds, small rodents, and many kinds of invertebrates. Prey is detected primarily by scent and is swallowed alive.

Reproduction Plains gartersnakes breed primarily in April or May, although possible fall mating has been reported. Several males may court and try to mate with the same female simultaneously. The females give birth at any time during the summer and into September, and the litter size can be highly variable (5–92) although it most commonly ranges from 10 to 20.

Predators and Defense Documented predators include raptors (hawks and kestrels), mammals (coyotes, foxes, mink, skunks, and house cats), and milk-

Plains gartersnakes, like this one from northwestern Indiana, usually have a dark body color.

The center stripe on Plains gartersnakes is bolder than the side stripes and often orange.

snakes. When threatened, their first response is to attempt a rapid escape. When captured or cornered they may strike at and bite their attacker. Other defensive displays include releasing musk, hiding the head beneath the body, and waving the tail from side to side.

Conservation Habitat loss and alteration for development or agriculture are major threats to this species, especially in Ohio and Wisconsin, where its geographic range is limited. Like other snake species that spend much of their time aboveground, Plains gartersnakes are highly susceptible to road mortality.

What's in a Name? This species was originally named *Eutania radix* in 1853 by S. F. Baird and C. Girard from a specimen collected in Racine, Wisconsin, by Dr. Philo Romayne Hoy. Dr. Hoy later became fish commissioner of Wisconsin and president of the Wisconsin Academy of Art, Science, and Letters. Baird and Girard did not explain their choice of *radix* for the species name, but *radix* means "root" in Latin, and the name may refer to the Root River, where the type specimen was collected.

COMMON GARTERSNAKE *Thamnophis sirtalis*

SCALES
Keeled

ANAL PLATE
Single

BODY SHAPE
Slender to moderately stocky

BODY PATTERN AND COLOR
Highly variable but typically with a dark body and either three stripes (usually yellowish) or a checkered appearance above and a solid pale white, yellow, or gray belly

DISTINCTIVE CHARACTERS
A dark vertical line borders each yellow lip scale

SIZE

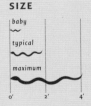

Description Common gartersnakes typically have a dark back with three yellow stripes that run the length of the body, but variation in the color and intensity of striping can occur both within local populations and among subspecies. The belly is usually a solid color, ranging from pale yellow to whitish or gray to greenish, with a black mark on the outer edge of each belly scale. The lip scales are yellowish or white and usually have conspicuous vertical dark markings between them but not as prominent as in the Plains gartersnake. Common gartersnakes have a distinctively colored red-and-black tongue. Four subspecies occur in the eastern United States. The eastern gartersnake (T. s. *sirtalis*) is highly variable

top Most common gartersnakes have light-colored stripes down their back.

A blue-striped gartersnake from Orange County, Florida

Some common gartersnakes have a checkerboard pattern with no obvious stripes.

Common gartersnake from Aiken County, South Carolina

COMMON GARTERSNAKE *Thamnophis sirtalis*

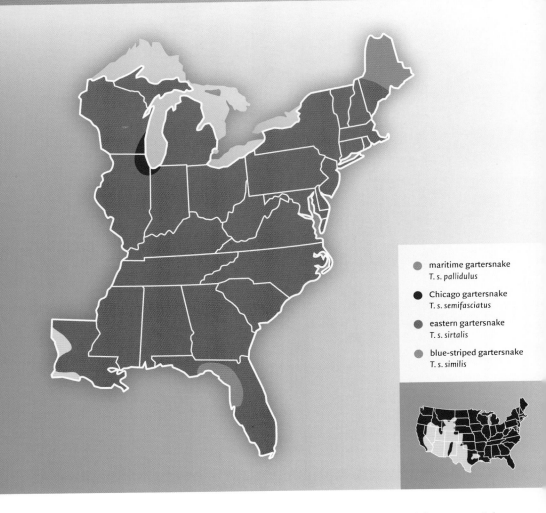

- maritime gartersnake *T. s. pallidulus*
- Chicago gartersnake *T. s. semifasciatus*
- eastern gartersnake *T. s. sirtalis*
- blue-striped gartersnake *T. s. similis*

in appearance, but most individuals have three yellow, brownish, or greenish stripes on a dark body. Some individuals, especially in Florida, have a vague or distinct checkerboard pattern that may obscure the stripes to some degree. The blue-striped gartersnake (*T. s. similis*) found in Florida has a dark brown or black body with a pale stripe down the center and a light blue stripe on each side. The Chicago gartersnake (*T. s. semifasciatus*) has yellowish stripes similar to the eastern subspecies, but large dark spots may interrupt the side stripes, and black bands are typically visible on the sides behind the neck region. The longitudinal stripes of the maritime gartersnake (*T. s. pallidulus*) are muted or

Baby common gartersnakes are born during summer or fall.

even absent, especially on the posterior part of the body, which is dull brownish, grayish, or olive. Dark spots create a checkerboard appearance in some individuals.

What Do the Babies Look Like? Baby gartersnakes look like the adults but with a darker head.

Distribution and Habitat Common gartersnakes are found throughout every eastern state except in some parts of western and southern Louisiana. Suitable habitats are generally near water and include the edges of drainage ditches, streams, swamps, ponds, wet meadows, and pastures. They prefer open, grassy habitats but are occasionally found in wooded areas. Common gartersnakes are among the most commonly seen snakes around bogs or streams at high elevations in the Appalachians and in many northern localities.

Behavior and Activity Common gartersnakes are noted for their tolerance of cool temperatures and are often active during the winter on warm or sunny days, where they can be seen basking on bushes, rocks, or grassy areas along streams or ponds. Many studies have shown that gartersnakes may travel more than a mile during their migration from feeding habitats to communal hibernation sites; in the Southeast and possibly other regions they may hibernate alone. They are active primarily during the day but occasionally forage at night during the summer in southern localities.

Food and Feeding Frogs, toads, and earthworms are the primary foods, but the diet may include many kinds of invertebrates, fish, small snakes, baby birds, mice, and shrews. Prey is detected primarily by scent and is generally swallowed alive. After capturing active prey

Common gartersnakes often search for small prey in grassy areas.

such as mice, common gartersnakes may chew vigorously, presumably so that their mildly toxic saliva will help to subdue the animal.

Reproduction Common gartersnakes mate primarily in the spring, although fall courtship has been observed and winter mating can occur in southern Florida. Females that are ready to mate release pheromones that attract males, sometimes many males, which may try to mate simultaneously with one female. Females can store sperm from previous matings for many months, and thus the offspring in a single litter may have multiple fathers. Young are born during the summer or fall, and litter size ranges from 1 to 101 (usually 20–30).

top Common gartersnakes have mildly toxic venom with which they subdue amphibian prey.

bottom Most gartersnakes do not hesitate to bite when threatened.

Predators and Defense Their long-distance movements, almost year-round activity, and association with both terrestrial and aquatic habitats expose gartersnakes to a broad spectrum of predators, including snakes, turtles, birds, mammals, spiders, frogs, and large fish. A gartersnake's first response when threatened is to try to escape. Some may feign death or hide the head beneath the body. Captured individuals may expand the head and body to appear larger and usually deliver open-mouthed strikes, eventually biting and chewing on the captor. They release a sweet-smelling, somewhat nauseating musk when first captured. The saliva has been reported to cause swelling and itching in some people.

Conservation In California, the San Francisco gartersnake (T. s. *tetrataenia*) is federally protected because so much of its habitat has been lost. Many populations throughout the eastern United States are being lost as a result of urbanization and development as well. Gartersnakes are frequent victims of highway mortality in many regions of the United States and Canada. However, they are also among the most commonly seen snakes in suburban areas in many parts of their range.

What's in a Name? Carl Linnaeus named this species *Coluber sirtalis* in 1758 based on a specimen from Canada provided by the Scandinavian natu-

Common gartersnakes are tolerant of cool temperatures and are often active on sunny days during winter or early spring.

ralist Pehr Kalm, who also wrote the first scientific description of Niagara Falls. The gartersnake was the fourth snake from North America to be given a formally accepted scientific name. This snake is often referred to as the eastern gartersnake, but because it is found from coast to coast and because three other species of gartersnakes are also found in the eastern United States, many herpetologists view that name as inappropriate. The scientific name *sirtalis* is of uncertain origin, although Kraig Adler of Cornell University suggested that Linnaeus combined the Greek word *siro*, meaning "cord," "rope," or "string" with the Latin word *talis*, meaning "such" or "the like," to describe the snake's ropelike appearance.

> **DID YOU KNOW?**
>
> The official state reptile of three eastern states is a snake: Massachusetts (common gartersnake), Ohio (racer), and West Virginia (timber rattlesnake). Only one western state, Arizona, has a snake (the ridge-nosed rattlesnake) as its state reptile.

A dozen subspecies have been proposed for this widespread, transcontinental species, four of which occur in the eastern states: the eastern gartersnake (*T. s. sirtalis*) described by Linnaeus; the Chicago gartersnake (*T. s. semifasciatus*) described by Edward Drinker Cope in 1892; the maritime gartersnake (*T. s. pallidulus*) described by G. M. Allen in 1899; and the blue-striped gartersnake (*T. s. similis*) described by Douglas A. Rossman in 1965. The derivations of the subspecific names include those from Latin *semi*, meaning "half," *fasciatus*, meaning "banded," and *pallidulus*, meaning "somewhat pale," all of which are descriptive of the particular subspecies. The word *similis* in Latin means "resembling," which refers to the similarity in appearance between the blue-striped gartersnake and blue-striped ribbonsnake (*T. s. nitae*) occurring in the same region.

BUTLER'S GARTERSNAKE Thamnophis butleri

SCALES
Keeled

ANAL PLATE
Single

BODY SHAPE
Moderately stocky

BODY PATTERN AND COLOR
Highly variable but typically with a dark body and either three stripes (usually yellowish) or a light brown or olive body color

DISTINCTIVE CHARACTERS
Head smaller than most other gartersnakes

SIZE
baby
typical
maximum

0' 2' 4'

Description Butler's gartersnakes are olive, brownish, or black with distinct longitudinal yellow stripes that may have an orange tinge. Small dark spots are often visible between the center and side stripes. The belly is solid greenish yellow with dark markings along the edge of each belly scale. The head is comparatively smaller than that of the common gartersnake. The lip scales are yellow and may have dark vertical lines that separate them.

What Do the Babies Look Like? Baby Butler's gartersnakes look like the adults but with comparatively larger heads.

top Butler's gartersnakes typically have three distinct yellow longitudinal stripes on a dark body.

BUTLER'S GARTERSNAKE *Thamnophis butleri*

Distribution and Habitat Butler's gartersnakes are found in scattered localities in four eastern states (Indiana, Ohio, Michigan, and Wisconsin), typically in moist, open grassy habitats or alongside streams. These snakes are noted for being found in urban and suburban parks and lawns. They appear to avoid forested habitats.

Behavior and Activity Butler's gartersnakes hibernate from October or November until March or April. They are most commonly seen during April and May after they emerge from hibernation. They are most active aboveground during the day but will also move around at dusk during warm periods.

The head of a Butler's gartersnake is smaller than that of a common gartersnake.

Food and Feeding Various species of earthworms are the predominant food in the wild, although diet records also include leeches and small frogs.

Reproduction Butler's gartersnakes mate upon emergence from hibernation in late March or early April. As in many other species of gartersnakes, several males may try to mate simultaneously with a single female. Litter size ranges from 4 to 20, with an average of 9 young being born in August or early September.

Predators and Defense These small snakes are presumably prey for many larger predators such as snakes, birds, and mammals, but few predatory records have been documented. Large crayfish have been suggested as a potential threat to some populations. Defensive behavior includes rapid wriggling in an attempt to escape and releasing musk. They rarely strike or bite when captured.

Conservation Feral cats are a source of mortality in suburban areas, and loss of the preferred habitat due to urbanization is a continued threat. This species is considered to be a species in need of protection in Indiana and Wisconsin.

What's in a Name? Edward Drinker Cope described this species in 1889 in the *Proceedings of the U.S. National Museum*, expressing his wish to "dedicate this handsome species" to Mr. A. W. Butler of Brookville, Indiana, who sent the first specimen known to science to the Smithsonian. Butler's gartersnake, the Plains gartersnake (T. *radix*), and the short-headed gartersnake (T. *brachystoma*) have been problematic for herpetologists for many years, sometimes being lumped together as the same species or considered to be closely related species that hybridize. Some authorities consider populations in southeastern Wisconsin and northeastern Illinois to be hybrids between T. *radix* and T. *butleri*.

EASTERN RIBBONSNAKE *Thamnophis sauritus*

SCALES
Keeled

ANAL PLATE
Single

BODY SHAPE
Very slender

BODY PATTERN AND COLOR
Dark back, usually with three yellow stripes above and a pale whitish or yellowish belly

DISTINCTIVE CHARACTERS
Tail proportionately longer than that of most other snakes; white or yellow spot in front of large eye

SIZE
baby
typical
maximum
0'   2'   4'

Description Eastern ribbonsnakes characteristically have three light, usually yellowish, stripes down the length of their dark brown or black body; the belly can be white or yellow. The lip scales are yellowish or white with no vertical dark markings between them. A small white mark is usually present in front of the eye. Four subspecies are recognized in the eastern United States, including the eastern ribbonsnake (T. s. sauritus), which matches the general description of the species. The stripe down the back of the peninsula ribbonsnake (T. s. sackenii) is either absent or less distinct than the side stripes. The blue-striped ribbonsnake (T. s. nitae) has blue stripes on the sides, and the middle stripe is absent or poorly defined. The northern ribbonsnake (T. s. septentrionalis) is usually black or dark brown with a center stripe that can appear brownish and side stripes that are pale yellow.

top Eastern ribbonsnakes have slender bodies with tails that are proportionately longer than those of most other snakes.

EASTERN RIBBONSNAKE *Thamnophis sauritus*

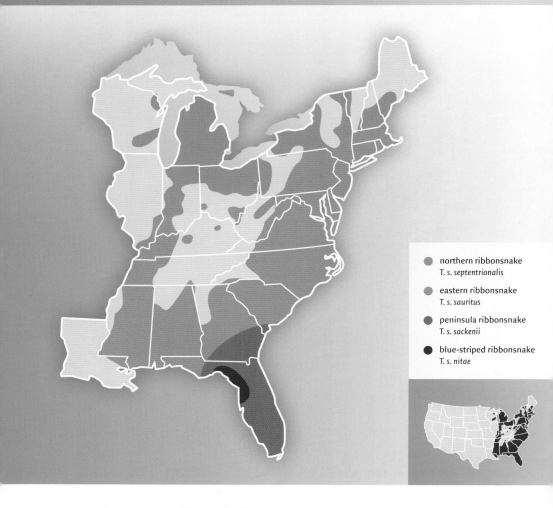

- northern ribbonsnake
  T. s. septentrionalis
- eastern ribbonsnake
  T. s. sauritus
- peninsula ribbonsnake
  T. s. sackenii
- blue-striped ribbonsnake
  T. s. nitae

What Do the Babies Look Like? Baby eastern ribbonsnakes look like the adults.

Distribution and Habitat One or more of the subspecies are found in part of every eastern state and in all or most of the majority of them. They are absent from large portions of Tennessee, Kentucky, West Virginia, and Pennsylvania, and only isolated populations occur in Illinois and Wisconsin. Eastern ribbonsnakes are more aquatic than most gartersnakes and are often abundant along the edges of permanent and semipermanent aquatic areas such as ponds, marshes, swamps, streams, and rivers. They frequently climb into low bushes overhanging or near the water.

Behavior and Activity Eastern ribbonsnakes hibernate during the winter in northern states but in more southerly locations may be found any season of the year if daily temperatures are warm or sunny conditions allow individuals to bask. They are active mostly during the day but may venture out in early evening or at night during hot summer periods.

Food and Feeding Eastern ribbonsnakes eat primarily amphibians, especially frogs and toads, which they usually swallow rear end first. They also feed on small fish, which they capture by swimming open-mouthed through the water. Spiders, caterpillars, and juvenile ribbonsnakes have also been reported in their diet. They search for and detect prey by both vision and smell, and swallow prey while it is still alive.

Reproduction Eastern ribbonsnakes mate in spring, and males apparently find females by following their pheromone trails. The females give birth in the summer or early fall to litters that usually range from 10 to 15 young. The largest litter size recorded for the eastern ribbonsnake is 26.

Predators and Defense Common predators include other snakes (especially kingsnakes and cottonmouths), wading birds, wetland mammals, bullfrogs, and larger fish. When encountered, ribbonsnakes always try to escape either on land or into the water. When captured they release a sick-

In contrast to most gartersnakes, ribbonsnakes have yellow or white lips with no vertical dark lines.

The peninsula ribbonsnake has yellow stripes on the sides but often has no central stripe.

Eastern ribbonsnake from Oakland County, Michigan

eningly sweet musk and sometimes bite. The end of the snake's long tail may break off if it is grabbed and will not grow back.

Conservation Eastern ribbonsnakes appear to be more vulnerable to disturbance than some other *Thamnophis* species. Because of their dependence on wetlands, eastern ribbonsnake populations may be affected by the destruction or alteration of aquatic habitats and their margins.

What's in a Name? Carl Linnaeus described and named this species *Coluber saurita* in 1766. In Latin *sauritus* means "lizard-like," probably referring to the long tail of the described specimen. In 1859 Robert Kennicott described *Eutaenia sackenii* as a new species named in honor of Russian diplomat and entomologist Baron C. R. Osten-Sacken, who collected the first specimen in Florida. It later became recognized as the subspecies T. s. *sackenii*. The blue-striped subspecies, T. s. *nitae*, was described in 1963 by Douglas A. Rossman, who named it in honor of his wife, Nita, who caught the first specimen. Rossman also described the northern ribbonsnake in 1963, naming it T. s. *septentrionalis* in reference to the Latin word *septentrio*, meaning "north."

WESTERN RIBBONSNAKE *Thamnophis proximus*

SCALES
Keeled

ANAL PLATE
Single

BODY SHAPE
Very slender

BODY PATTERN AND COLOR
Dark back with three distinct yellow or orange stripes above and a pale whitish or yellowish belly

DISTINCTIVE CHARACTERS
Bold center stripe of gold or orange; yellow or white spot in front of eye

SIZE
baby
typical
maximum
0'   2'   4'

Description Western ribbonsnakes have three distinct yellow stripes down the length of the black or dark brown body, the center stripe often being more orange; the belly can be white or yellow. The lip scales are yellowish or white with no vertical dark markings between them. A small white or yellowish mark is present in front of the eye. Two subspecies occur in the eastern states. The western ribbonsnake (T. p. proximus) typically has distinct yellow or greenish yellow stripes, and the Gulf Coast ribbonsnake (T. p. orarius) is characterized by a wide, gold central stripe. Intergradation occurs between the two subspecies of western ribbonsnakes in Louisiana (see map).

top Most western ribbonsnakes have a bold center stripe of gold or orange. All ribbonsnakes have a yellow or white spot in front of the eye.

The longitudinal stripes of a western ribbonsnake can be greenish.

Western ribbonsnakes are born in summer or early fall.

What Do the Babies Look Like? Baby western ribbonsnakes are miniatures of the adults, but like many baby snakes their eyes may appear disproportionately large.

Distribution and Habitat The western ribbonsnake subspecies (T. p. proximus) is found in all six eastern states that are adjacent to the Mississippi River, from Louisiana to Wisconsin as well as in northwestern Indiana. The Gulf Coast ribbonsnake is found in southern Louisiana. This species is nearly always associated with aquatic habitats and can be abundant along the edges of permanent and semipermanent wetlands, including lakes, ponds, open marshes, swamps, streams, and rivers. They often climb into low shrubs or bushes overhanging or near the water.

Behavior and Activity Western ribbonsnakes hibernate in underground retreats during winter in the northern parts of their range but in southern areas may be found year-round during warm periods resting openly in the sun on low vegetation on cool, sunny days or in more shaded spots during warm days. They are primarily active during the day for most of the year but during hot summer periods may be out at dusk or later in the evening.

Food and Feeding Western ribbonsnakes eat adult frogs and toads as well as tadpoles, small fish, and salamanders. They also have been reported to eat lizards and even other ribbonsnakes.

Reproduction Western ribbonsnakes mate in early spring, immediately after hibernation in northern populations. The females give birth anytime during the summer and early fall, with litters being recorded from June to October. Litter

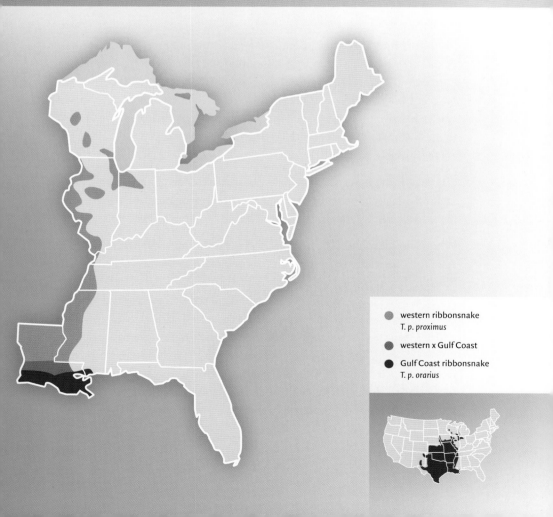

WESTERN RIBBONSNAKE *Thamnophis proximus*

- western ribbonsnake *T. p. proximus*
- western x Gulf Coast
- Gulf Coast ribbonsnake *T. p. orarius*

Western ribbonsnakes live near aquatic habitats but spend much of their time on land.

size usually ranges from 10 to 15 young. The largest litter size recorded for the western ribbonsnake is 36.

Predators and Defense Common predators documented for this species include common kingsnakes, cottonmouths, and larger western ribbonsnakes. Carnivorous mammals, wading birds, bullfrogs, fish, and aquatic beetles, which may feed on juveniles, also take their toll. When encountered, ribbonsnakes always try to escape either on land or into the water, where they usually dive to the bottom and hide beneath vegetation. When captured they release musk and occasionally will bite.

Conservation Western ribbonsnake populations depend both on aquatic habitats and on the surrounding terrain, and therefore can be affected detrimentally by the destruction of wetlands as well as the adjacent upland habitat.

What's in a Name? Thomas Say described this species in 1823 based on a specimen he collected as the chief zoologist on the James Expedition from Pittsburgh to the Rocky Mountains. *Proximus* means "near or nearest" in Latin, perhaps a reference to the western ribbonsnake's close resemblance to the eastern ribbonsnake (*T. sauritus*) and two species of gartersnakes (*T. ordinoides* and *T. s. parietalis*). The name of the Gulf Coast ribbonsnake subspecies (*T. p. orarius*) described by Douglas A. Rossman in 1963 based on his dissertation from the University of Florida is derived from the Latin word *orarius*, meaning "of the coast" and referring to the snake's geographic distribution. The species was recognized for many years as a subspecies of *T. sauritus*.

EASTERN HOGNOSE SNAKE Heterodon platirhinos

SCALES
Keeled

ANAL PLATE
Divided

BODY SHAPE
Heavy bodied

BODY PATTERN AND COLOR
Highly variable; sometimes blotched with different colors, including red, yellow, orange, or gray; sometimes solid gray or black

DISTINCTIVE CHARACTERS
Prominent, pointed nose; spectacular threat display

SIZE
baby
typical
maximum
0' 2' 4'

Description Eastern hognose snakes usually have a blotched pattern with color combinations that can vary greatly within a local population or even among siblings within the same clutch of eggs. The colors on the back and sides can include shades of red, orange, yellow, gray, olive, brown, or black. In some areas of the Southeast most individuals become solid black or dark gray above as adults. The belly is also variable, ranging from pale gray to dark and sometimes having lighter or darker blotches. The belly is

top Eastern hognose snakes have robust bodies and are found throughout most eastern states.

The nose of juvenile eastern hognose snakes is not quite as pointed as that of the adults.

EASTERN HOGNOSE SNAKE *Heterodon platirhinos*

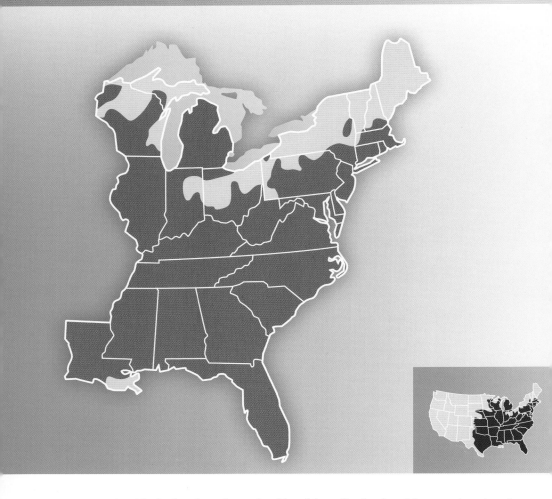

noticeably darker than the underside of the tail. The tip of the nose is pointed but not as noticeably turned up as that of the southern hognose snake. The characteristic bluffing and death-feigning displays (see below) are so dramatic that they should be considered part of the general description of the species.

What Do the Babies Look Like? Baby hognose snakes, even those that will turn solid black or gray as adults, always have a blotched pattern.

Distribution and Habitat Eastern hognose snakes occur throughout all or most parts of every eastern state except Vermont and Maine. They inhabit a wide variety of habitats, including abandoned agricultural fields, open pine forests,

A boldly patterned eastern hognose snake

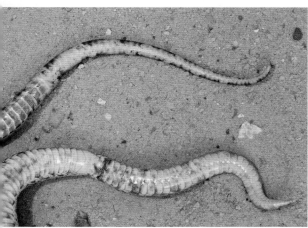

The noticeably darker belly of an eastern hognose snake (*top*) contrasts with the lighter colored underside of the tail. In the southern hognose snake (*bottom*) belly and tail are similar in color.

and rocky forested hillsides. They usually are associated with loose soil, and if living in forests usually stay near the edge where they can venture into more open areas. Hognose snakes characteristically burrow into sandy soils using their pointed, spadelike nose as a shovel.

Behavior and Activity Aboveground activity generally begins early in the spring and ends late in the fall. Recently hatched juveniles are commonly found in September and October, and adults may bask on sunny days during the winter months in the warmest parts of the range. Eastern hognose snakes are active aboveground only during the daytime.

Part of the defensive behavior of eastern hognose snakes includes playing dead.

Food and Feeding Toads along with an occasional frog or salamander make up most of the diet. A few records exist of wild-caught individuals having eaten small mud turtles, lizards, small snakes, birds, and small mammals as well as a variety of invertebrate prey. The snake may use its pointed nose to dislodge toads buried in loose soil or sand. Toads inflate their body as a defense mechanism when captured by snakes, but eastern hognose snakes use their enlarged posterior teeth to "pop" and deflate the toads, thus making them easier to swallow. Enzymes in the snake's digestive system neutralize the skin toxins of toads.

Reproduction Breeding occurs primarily in the spring but also may occur during the fall. Eggs are laid anytime from May through August, and clutches typically number about 20–25 eggs (range, 4–61). Eggs are usually laid under rocks or in loose soil. Some nests have been found in sawdust piles. The eggs hatch in about 6–7 weeks.

Eastern hognose snakes spread their neck when threatened. Adult eastern hognose snakes are often solid black in some regions.

Predators and Defense Natural predators of eastern hognose snakes are numerous, in part because the snakes are active aboveground in the daytime and occur in a wide array of habitats. Predators include other snakes, birds of prey, and carnivorous mammals. Hognose snakes have a distinctive repertoire of responses to threats from potential predators, including humans. When initially threatened, the snake spreads its neck, which makes the snake appear larger. Some individuals hiss, open the mouth, and move the head forward in a striking motion, but without actually trying to bite. If this formidable display does not intimidate or discourage the attacker, the snake will appear to go into convulsions and roll over on its back as if dead, often disgorging a recently eaten meal as part of the display. The tongue hangs out of the snake's open mouth,

left Toads are the most common prey of eastern hognose snakes.   *right* The enlarged posterior teeth of eastern hognose snakes are used to deflate toads that inflate their bodies.

and capillaries in the mouth may rupture, producing a large amount of blood to add to the effect. The snake remains on its back with open mouth during the death-feigning display and will actually roll back over if it is righted. Eastern hognose snakes will not intentionally bite people, although injuries may occur accidentally if a person catches a finger on one of the snake's teeth when the snake's mouth is held open. The saliva is mildly venomous and sometimes causes an allergic reaction on the rare occasions when someone is bitten. Unfortunately, people sometimes mistake harmless hognose snakes with blotched patterns of red or orange for copperheads and kill them.

Conservation   Hognose snakes commonly travel overland and frequently become highway mortality victims. Although they are terrestrial, their dependence on toads as prey makes the protection of small wetlands where toads breed critical in some areas.

What's in a Name?   This species was described in 1801 by Pierre André Latreille, who was considered to be the foremost entomologist of his time. He was also a respected taxonomist and expert in zoological nomenclature who coauthored the classic *Histoire Naturelle des Reptiles* (*Natural History of Reptiles*) with C. S. Sonnini. In Greek the word *hetero* means "different" and *odontos* means "tooth," the combined generic name referring to the contrast between the enlarged rear teeth and relatively smaller front teeth. The species name is derived from the Greek words *plati*, meaning "flat," and *rhinos*, meaning "nose," describing the characteristic flattened, pointed nose. Hognose snakes are commonly called spreading adders or puff adders in many areas.

WESTERN HOGNOSE SNAKE Heterodon nasicus

SCALES
Keeled

ANAL PLATE
Divided

BODY SHAPE
Heavy bodied

BODY PATTERN AND COLOR
Light brown or tan with large, dark brown blotches down center and on sides; belly black with patches of yellow or white

DISTINCTIVE CHARACTERS
Prominent, upturned nose; dramatic threat display; black belly

SIZE
baby
typical
maximum
0'   2'   4'

Description Western hognose snakes are light brown or olive with large dark blotches down the center of the back and tail, and smaller blotches on the sides. Alternating dark and light bands extend from the top of the head and eye to the mouth, giving the face a striped appearance alarmingly similar to that of various venomous rattlesnakes in the western and southwestern portions of the geographic range. The belly is predominantly black interspersed with patches of yellow or white. The distinctive bluffing, mock-striking, and death-feigning displays are so remarkable that they qualify as part of the general description of the species.

top Western hognose snakes typically have light brown bodies with dark blotches.

Western hognose snakes have an upturned nose.

WESTERN HOGNOSE SNAKE *Heterodon nasicus*

What Do the Babies Look Like? Baby western hognose snakes are similar in pattern to the adults but more brightly colored.

Distribution and Habitat East of the Mississippi River, western hognose snakes occur naturally only in western Illinois. They have been introduced into Kankakee and Iroquois counties south of Chicago where they are not native, but viable populations may now exist. They are found in a diversity of habitats, especially in sandy soils of open prairielands, pastures, and abandoned agricultural fields. These snakes routinely excavate loose, sandy soils with the large upturned nose in search of prey or to create temporary burrows.

Behavior and Activity Activity in Illinois generally begins in mid-to-late spring and ends in the fall when temperatures cool. Western hognose snakes are active aboveground primarily during the daytime in Illinois.

Food and Feeding Although they specialize on toads, western hognose snakes have a more diverse diet than the other two species of hognose snakes, the array of prey in the wild including amphibians, reptiles, birds, and mammals as well as frogs, salamanders, lizards, snakes, and bird eggs. The spadelike nose is commonly used to unearth buried prey, including reptile eggs. They have been documented eating the eggs of at least five species of turtles. The enlarged posterior teeth are used to puncture toads that inflate their body as a defense mechanism when captured.

A black belly is characteristic of western hognose snakes. This individual is playing dead.

Reproduction Western hognose snakes mate mostly in the spring from March into May, but some may breed in August. An average of 10–11 eggs (2–24) are laid in sandy or other loose soil that the female can dislodge with her nose. The eggs are laid in June or July and hatch in August or early September.

Predators and Defense Natural predators include hawks, crows, and coyotes. Other snakes, including massasaugas, have also been documented to eat this species in western parts of its range, and cannibalism has been reported. When encountering a person or potential predator, a threatened western hog-

Western hognose snakes use their pointed nose to burrow beneath the surface in search of prey.

nose snake spreads its neck and may hiss and make a close-mouthed strike but without actually trying to bite. Next the snake will pretend to die, convulsing and rolling over on its back. If food is in its stomach, the snake will usually regurgitate it and sometimes will bleed from the mouth. Like the other hognose snakes, it will remain on its back during the death-feigning display and will flip itself back over if placed upright. Their similarity in body pattern and head striping to rattlesnakes found in the same habitats may constitute an early passive defense and make them appear more menacing to some would-be predators. Unfortunately, even though the harmless western hognose and venomous massasauga do not occur in the same localities in Illinois, humans often mistake the harmless snake for the venomous one and kill it.

Conservation Like other species of snakes that frequently travel overland, western hognose snakes suffer high levels of mortality on roads. The species is protected by the state of Illinois.

What's in a Name? This species was initially described in a single sentence by S. F. Baird and C. Girard in 1852 based on a specimen from Texas provided to the Museum of the Smithsonian Institution. The Latin word *nasica* means "nose" and refers to the distinctive turned-up nose.

SOUTHERN HOGNOSE SNAKE *Heterodon simus*

SCALES
Keeled

ANAL PLATE
Divided

BODY SHAPE
Very stout

BODY PATTERN AND COLOR
Light brown or sometimes reddish body; dark brown blotches down the back and smaller blotches on the sides

DISTINCTIVE CHARACTERS
Sharply upturned nose

SIZE
baby
typical
maximum
0'   2'   4'

Description This is the stoutest snake in the eastern United States. Although southern hognose snakes are relatively short, adults weigh more than twice as much as many other species of snakes of the same length. The end of the nose turns upward at a sharp angle and is much more upturned and spadelike than that of the eastern hognose. The body is usually light brown or reddish brown with dark brown blotches on the back that alternate with smaller ones on the sides. The belly and underside of the tail are light brown or gray. Southern hognose snakes are never solid gray or black like eastern hognose snakes. The threat and death-feigning displays of the southern hognose snake are less dramatic than those of the eastern hognose snake but are nonetheless characteristic of the species.

top Southern hognose snakes always have dark brown blotches on a lighter colored body.

What Do the Babies Look Like? The babies are miniatures of the adults.

Distribution and Habitat Southern hognose snakes historically occurred in parts of the Coastal Plain of southern North Carolina, South Carolina, most of Florida, Georgia, and Alabama into southern Mississippi. No specimens have been documented from Alabama or Mississippi since the 1970s, and the species is assumed to be extirpated from those states. These snakes favor dry, well-drained, sandy soils and can be found in longleaf pine forests, abandoned agricultural fields, and a variety of scrubby oak woodlands. They are sometimes found in coastal sand dunes and in cultivated fields.

Behavior and Activity Southern hognose snakes may be found aboveground in every month, including sunny days in winter, and are active only during the daytime hours. They spend much of their time underground, burrowing through the sandy soil using the upturned snout as a shovel. A high proportion of the southern hognose snakes captured are found while crossing roads.

Southern hognose snakes have a more flattened, spade-like nose than eastern hognose snakes.

Food and Feeding Like their larger relative, southern hognose snakes eat primarily toads, including spadefoot toads. They use the upturned snout to dig up toads buried in loose

This southern hognose snake from Baker County, Georgia, has a reddish tinge.

SOUTHERN HOGNOSE SNAKE *Heterodon simus*

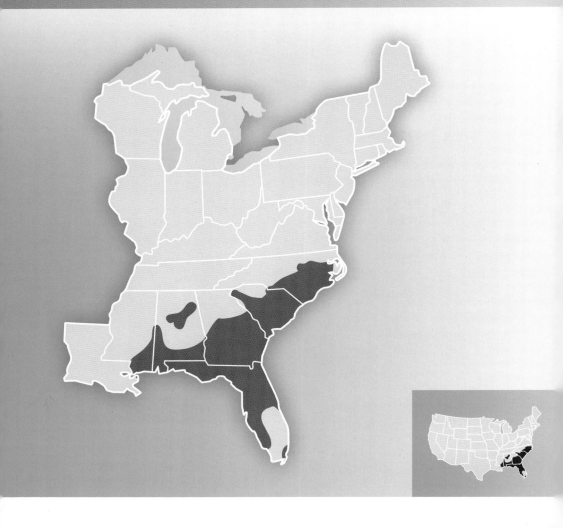

soil, and if the toad expands its body in defense they use their enlarged posterior teeth to puncture it. Southern hognose snakes sometimes feed on frogs, small lizards, baby mice, and insects. They may subdue prey with their toxic saliva.

Reproduction Little is known regarding many aspects of the reproductive biology of southern hognose snakes, but courtship or mating has been observed in spring (May) and fall (September–November). Six to 19 (average about 9) white, elongated eggs are laid in mid-to-late summer, and the young typically hatch in September or October.

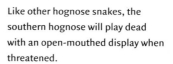

Like other hognose snakes, the southern hognose will play dead with an open-mouthed display when threatened.

Southern hognose snakes often cross highways on cool, sunny days in early autumn.

Predators and Defense Natural predators include various snakes, birds, and mammals associated with sandhill habitats. Nonnative fire ants, which also are found in sandy areas, may attack the eggs and hatchlings. The southern hognose snake's response to a predatory threat, which includes spreading the neck and playing dead, is similar to that of the eastern hognose but somewhat less dramatic.

Conservation The southern hognose snake has been extirpated or is in danger of extinction throughout much of its geographic range and has been petitioned for federal listing under the Endangered Species Act. No specimens have been found in Alabama or Mississippi since the 1970s in areas where they once occurred, and they have disappeared from two-thirds of the counties where they were formerly found in the Carolinas, Georgia, and Florida. Nonnative fire ants have been implicated in their decline, but the steady loss and fragmentation of upland sandhill habitats through development is unquestionably a problem for the species in all areas. Many are killed while crossing roads adjacent to their habitat.

What's in a Name? Carl Linnaeus named this snake *Coluber simus* in 1766, based on a specimen he received from Alexander Garden of Charleston. Garden was a naturalist for whom the gardenia flower was named. The species name of *Heterodon simus* is derived from the Latin word *simulus*, meaning "flat-nosed" or "snub-nosed," referring to the distinctive upturned snout. Like the eastern hognose, this species is called spreading adder or puff adder in some regions.

MOLE & PRAIRIE KINGSNAKES *Lampropeltis calligaster*

SCALES
Smooth

ANAL PLATE
Single

BODY SHAPE
Moderately robust to robust

BODY PATTERN AND COLOR
Variable from solid brown to brownish gray with dark brown to reddish blotches

SIZE

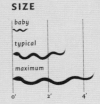

Description Mole and prairie kingsnakes have a rounded head, moderately robust body, and relatively short tail. All three subspecies (mole kingsnake, *L. c. rhombomaculata*; prairie kingsnake, *L. c. calligaster*; and south Florida mole kingsnake, *L. c. occipitolineata*) typically have a light brown body with reddish or brownish blotches down the back. The blotches on some mole kingsnakes become very faded with age, resulting in a nearly solid brown snake. The belly is lighter than the back and often has a faint but darker blotching pattern. The three subspecies are distinguished by differences in numbers of blotches and scale counts. Intergradation occurs in the zone of contact between the mole and prairie kingsnakes (see map).

top Many mole kingsnakes, like this one from Coweta County, Georgia, have brown bodies with reddish blotches.

MOLE AND PRAIRIE KINGSNAKES *Lampropeltis calligaster*

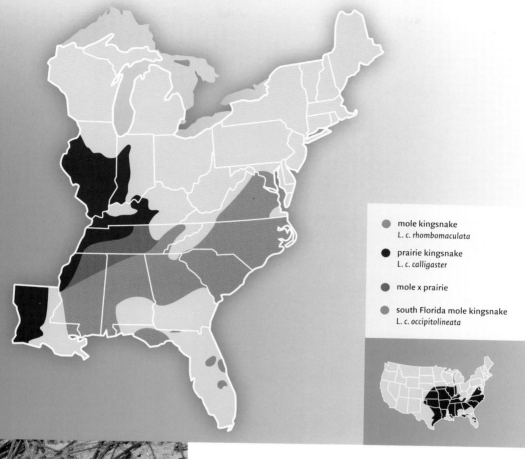

- mole kingsnake
  L. c. rhombomaculata
- prairie kingsnake
  L. c. calligaster
- mole x prairie
- south Florida mole kingsnake
  L. c. occipitolineata

Juvenile mole and prairie kingsnakes are more boldly patterned than older adults.

What Do the Babies Look Like? Babies are more boldly marked and have distinctive reddish blotches on the back.

Distribution and Habitat Prairie or mole kingsnakes occur in 14 of the eastern states. However, they are sporadically distributed in Florida and absent from most of the southern portions of South Carolina, Georgia, Alabama, and Louisiana, and from most of the Northeast and northern midwestern states.

above The body of some prairie kingsnakes, like this one from Vigo County, Indiana, can be grayish brown with darker blotches.

left The blotches on the body of mole kingsnakes are often red.

They live in a variety of habitats but are most abundant in open areas such as old agricultural fields, grasslands, savannas, and the edges of pastures. They can frequently be found beneath debris such as old boards and discarded sheets of tin strewn around old barns and abandoned houses.

Behavior and Activity Mole and prairie kingsnakes are typically active from spring to fall in cooler regions. They hibernate through the winter in most of their range but may be active on warm days in the southern part of the range, especially following afternoon thunderstorms. They presumably spend much of the time underground in rodent burrows but are active aboveground during the day in the cooler parts of the year; during the summer months they are more likely to be found in early evening or at night.

Food and Feeding Prairie and mole kingsnakes eat shrews, mice, moles, pocket gophers, voles, racerunner lizards, skinks, and fence lizards, which they kill by constriction. They sometimes eat frogs, other snakes, and bird eggs. They have been reported to eat small pit vipers such as copperheads and, like their larger relatives the common kingsnakes, are presumably immune to the effects of pit viper venom.

Reproduction The breeding season begins in spring and lasts through June. During mating, the male may bite the female on the neck. As many as 20 eggs (typically 8–14) are laid during summer. Mole kingsnakes usually lay smaller eggs than do prairie kingsnakes. The eggs hatch after an incubation period of about 2 months.

Predators and Defense Common kingsnakes, large hawks, and carnivorous mammals prey on mole and prairie kingsnakes. When captured these snakes may vibrate their tail, release musk, and bite. They usually settle down quickly and ordinarily are docile if handled gently, rarely biting humans.

Conservation The species is uncommon throughout much of its range, and in some cases it is difficult to know whether a population has been eliminated from a particular area or was never present. These secretive snakes are frequently killed on highways in agricultural areas and are probably victims of soil tillers and similar ground-breaking agricultural equipment.

What's in a Name? This species was described in 1827 by Richard Harlan from specimens in the Philadelphia Museum. All of the kingsnakes have smooth, shiny scales as alluded to by the Greek words *lampros*, meaning "shiny," and *pelta*, meaning "shield." The mole kingsnake (*L. c. rhombomaculata*) was recognized as a distinct subspecies by John Edwards Holbrook in 1840. R. M. Price described the south Florida mole kingsnake (*L. c. occipitolineata*) as an additional subspecies in 1987 and credited Ross Allen of Ross Allen's Reptile Institute for verifying the specimens.

Prairie kingsnake from southwestern Indiana

MILKSNAKE *Lampropeltis triangulum*

SCALES
Smooth

ANAL PLATE
Single

BODY SHAPE
Slender to moderately robust

BODY PATTERN AND COLOR
Louisiana milksnake: ringed completely around the body with red, yellow, and black; eastern milksnake and red milksnake: light colored with darker reddish or brown blotches down the back and on the sides

DISTINCTIVE CHARACTERS
Louisiana milksnake: black nose and red, yellow, and black body rings, with red being adjacent to black; eastern and red milksnakes: light colored with darker blotches down the back

SIZE
baby
typical
maximum
0'   2'   4'

Description Milksnakes are relatively robust, although smaller males may be more slender than females. The three eastern subspecies of milksnakes vary more dramatically from each other in appearance than some snake species vary from each other. The Louisiana milksnake (*L. t. amaura*) characteristically has brightly colored rings of red, yellow, and black completely encircling the body, resembling the scarlet kingsnake but with a black head. The eastern milksnake

top Typical eastern milksnakes have light brown bodies with dark blotches down the back.

(*L. t. triangulum*) is light colored, ranging from gray to light brown, with dark red or reddish brown blotches down the back and on the sides. This subspecies is characterized throughout much of its range by a light-colored V- or Y-shaped mark on the neck with the wide end pointing forward. The red milksnake (*L. t. syspila*) has a gray or whitish body with large red blotches down the back. The eastern milksnake and red milksnake hybridize with the scarlet kingsnake in areas of Tennessee and adjacent states. Despite the formal designation of many subspecies names for milksnakes, the variability in color and body patterns is extensive within the species, especially within broad zones of intergradation.

What Do the Babies Look Like? Young milksnakes look like the adults, except that the red blotches on baby eastern milksnakes are usually more brightly colored than those on adults.

right A milksnake from Brown County, Indiana

below Red milksnakes often have large, bright red blotches down the back that are edged with black.

MILKSNAKE *Lampropeltis triangulum*

- eastern milksnake
  L. t. *triangulum*
- red milksnake
  L. t. *syspila*
- Louisiana milksnake
  L. t. *amaura*

Distribution and Habitat Milksnakes are found in at least a small part of every eastern state except Florida. Louisiana milksnakes are most abundant in forests, where they often hide beneath the loose bark of dead trees, especially pines. Eastern and red milksnakes can be found in dense forest habitat but also occur in more open field habitats in some situations. They can frequently be found under rocks and debris in abandoned agricultural areas and grassy meadows, and under logs in and along the edges of woodlands.

Behavior and Activity This species is active in all warm months and in the southern part of its range may be active during warm periods in winter. Individuals can be found on the surface both night and day but are more active at night in warm regions in the Southeast. Milksnakes in the northern part of their range have been known to congregate at winter hibernation sites in barns, at the base of stone walls, and in the basements of occupied houses.

Food and Feeding Milksnakes feed on both vertebrates and invertebrates as well as the eggs of lizards, snakes, and birds. Documented prey includes several species of small rodents and shrews, small birds, amphibians, a dozen species of lizards, and numerous species of snakes, including copperheads and rattlesnakes. They are presumably immune to the effects of pit viper venom. Invertebrate diet records include slugs, beetles, and roaches. Prey may be detected by smell and actively hunted or ambushed, and is killed by constriction if necessary.

Reproduction The mating season lasts from early April into June. Some individuals mate in the fall, and the female stores the sperm until fertilization occurs the following spring. Females leave pheromone trails that males follow, and during courtship the male bites the neck of the female and rhythmically moves his body against hers. Eggs are laid during early summer inside rotting logs, under flat rocks, beneath moist bark, or in piles of rotting vegetation. Eastern milksnakes lay clutches of 1–24 eggs (average about 9); the Louisiana milksnake has a smaller clutch size, usually only 4 or 5 eggs. Milksnake eggs hatch after about 8 weeks in late summer or early fall.

Predators and Defense The primary predators of milksnakes are snake-eating snakes such as common kingsnakes and racers, birds of prey, and carnivorous mammals. Louisiana milksnakes resemble eastern coral snakes with their red, yellow, and black rings and black snout, and this resemblance may afford pro-

Young eastern milksnakes are more brightly colored than the adults.

Milksnakes subdue and kill lizards and other prey by constricting them.

tective mimicry that makes some predators, especially birds, more cautious about attacking them. Milksnakes usually vibrate the tail, release musk, and often bite and chew when captured.

Conservation As with many other species associated with woodland habitats, both residential and commercial development can be detrimental to this species. Milksnakes are frequently killed on roads.

What's in a Name? Although more than two dozen subspecies have been named, the original species description credited to B. G. E. Lacepède in 1789 was of the eastern milksnake subspecies (L. t. *triangulum*). The Greek words *tri*, meaning "three," and *angulum*, meaning "angle," presumably refer to the triangular pattern on the head characteristic of individuals that belong to this subspecies. The Louisiana milksnake (L. t. *amaura*) was described and named by Edward Drinker Cope in 1860 as a full species from a preserved specimen of unknown locality in the Smithsonian Institution Museum. Despite the vivid red, yellow, and black coloration of this species in life, Cope stated that the specimen had "no yellow marking whatever upon it," which presumably is what led to his choice of the subspecies name from Greek *amauros*, meaning "dark" or "dim." Cope described the red milksnake (L. t. *syspila*) in 1889 as a subspecies. The Greek word *sys* means "together," and *spilos* means "spots," presumably a reference to the "median black spots" on the belly mentioned by Cope. Although many herpetological authorities have long considered the closely related scarlet kingsnake (*Lampropeltis elapsoides*) a subspecies of the milksnake some now deem it a separate species.

SCARLET KINGSNAKE *Lampropeltis elapsoides*

SCALES
Smooth

ANAL PLATE
Single

BODY SHAPE
Slender to moderately robust

BODY PATTERN AND COLOR
Ringed completely around the body with red, yellow, and black; red nose

DISTINCTIVE CHARACTERS
Red nose; red and black body rings adjacent

SIZE

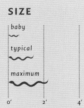

Description Scarlet kingsnakes are relatively robust, although smaller males may be less stout than females. They characteristically have brightly colored rings of red, yellow, and black completely encircling the body. The yellow bands may appear nearly white, and in some areas may appear more orange. A mixing of color patterns occurs in zones of hybridization between the scarlet kingsnake and both the eastern milksnake and the red milksnake. Most of these "hybrids" are recognized by authorities as milksnakes (i.e., L. triangulum).

What Do the Babies Look Like? The young look like the adults, although the yellow bands may be whitish.

top The classic color pattern of a scarlet kingsnake is a red nose with black rings that alternate between red and yellow rings.

Distribution and Habitat Scarlet kingsnakes are found in all or part of the Atlantic and Gulf coastal states from New Jersey to Louisiana and in Tennessee and Kentucky. They are most abundant in forests, where they often hide beneath the loose bark of dead trees, especially pines.

Behavior and Activity Scarlet kingsnakes are active in all warm months throughout their range and in the winter months during warm periods, but activity aboveground and the probability of seeing them are very low. Nonetheless, individuals can be found on the surface during both night and day and are decidedly more active at night during warm periods. They are noted for hiding beneath the loose bark of dead pine trees, both upright and fallen, and beneath ground cover, especially during early spring and autumn.

Food and Feeding Scarlet kingsnakes feed mostly on lizards, but small rodents, other snakes, eggs, fish, large insects, and earthworms have also been reported in the diet. Prey is killed by constriction.

Reproduction The mating season lasts from early April through May. Some individuals may also mate in the fall. Eggs are laid during early summer inside

left Lizards are a major prey item of scarlet kingsnakes.

below Scarlet kingsnakes commonly hide under tree bark or in holes in trees.

SCARLET KINGSNAKE *Lampropeltis elapsoides*

rotting logs, under moist bark, and in piles of decaying vegetation. Scarlet kingsnakes usually lay four or five eggs that hatch in late summer or early fall.

Predators and Defense The primary predators are snake-eating snakes, forest-dwelling birds of prey, and carnivorous mammals. Scarlet kingsnakes superficially resemble eastern coral snakes with their red, yellow, and black rings, and some herpetologists consider this resemblance to be a form of protective mimicry that makes some predators, especially birds, hesitate before attacking. Scarlet kingsnakes usually vibrate the tail, release musk, and often bite and chew

above Scarlet kingsnakes are often associated with pine forests.

left A colorful but oddly patterned scarlet kingsnake

when captured. They have been reported to be immune to the venom of pit vipers, but researchers at the Savannah River Ecology Laboratory observed an incident in which a scarlet kingsnake died within minutes of being bitten by a coral snake.

Conservation The conversion of woodland habitats to urban areas is harming this forest-dwelling species throughout its geographic range. At one time, scarlet kingsnakes were collected in large numbers for the pet trade, and according to some sources this practice continues in some parts of Florida. However, widespread captive-breeding programs and the resulting variety of kingsnakes and milksnakes now available to pet owners have likely reduced the removal of individuals from wild populations.

What's in a Name? This species was the last snake described in volume 2 of *North American Herpetology* (1838) by John Edwards Holbrook, who gave South Carolina and Georgia as the known range. The Latin words *elaps*, meaning "serpent," and *oides*, meaning "similar to," refer to this snake's superficial resemblance to the coral snake (*Micrurus fulvius*), which at one time was classified in the genus *Elaps*. Most authorities consider the scarlet kingsnake to be a separate species from the milksnakes (*Lampropeltis triangulum*), but some herpetologists may still consider them a subspecies of milksnake.

large terrestrial snakes

previous page Yellow ratsnake

COMMON KINGSNAKE *Lampropeltis getula*

SCALES
Smooth

ANAL PLATE
Single

BODY SHAPE
Robust; rounded head; relatively short tail

BODY PATTERN AND COLOR
Shiny black with white or yellow speckles or rings

DISTINCTIVE CHARACTERS
Rounded head; scales usually shiny

SIZE

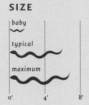

Description The back is typically black or dark brown with yellow or white markings (bands or spots) that vary regionally, by subspecies, or both; the belly is usually shiny black or mottled with yellow or white. The commonly recognized kingsnake subspecies in the eastern states are the eastern kingsnake (*L. g. getula*), with thin white or yellow bands across the back; the speckled kingsnake (*L. g. holbrooki*), with a yellow spot on most or all of the black body scales; and the eastern black kingsnake (*L. g. niger*), with limited spotting on the sides and banding absent or noticeable only as faint spotting. The Florida kingsnake (*L. g. floridana*) has scales that are predominantly yellowish brown or cream colored with a faint banding pattern visible along the body; the belly is light colored

top The eastern kingsnake has white or yellow bands across its back.

COMMON KINGSNAKE *Lampropeltis getula*

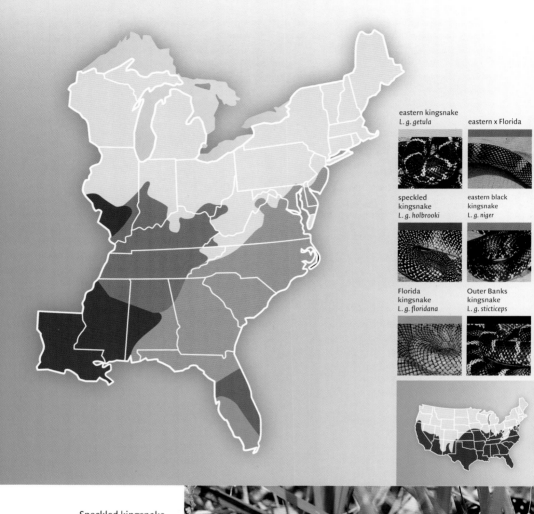

eastern kingsnake
L. g. getula

eastern x Florida

speckled kingsnake
L. g. holbrooki

eastern black kingsnake
L. g. niger

Florida kingsnake
L. g. floridana

Outer Banks kingsnake
L. g. sticticeps

Speckled kingsnake swallowing a racer

The speckled kingsnake has a yellow spot on most or all of the shiny black body scales.

left Some herpetologists recognize the Apalachicola kingsnake in Florida as a distinct subspecies. *right* The Florida kingsnake has a predominantly yellowish brown or cream colored body with a faint banding pattern.

with slightly darker spots. Eastern kingsnakes and Florida kingsnakes interbreed in some regions of central Florida, producing snakes that have variable color patterns. Some herpetologists recognize the Outer Banks (North Carolina) kingsnake (*L. g. sticticeps*), with its more brownish appearance and yellow spotting between the bands, and the profusely yellow-spotted and sometimes banded Apalachicola kingsnake as subspecies. Intergrades occur between subspecies in areas of contact.

An eastern kingsnake eating a cottonmouth

> **DID YOU KNOW?**
>
> Common kingsnakes are immune to the venom of pit vipers such as copperheads, cottonmouths, and rattlesnakes. Whether kingsnakes are immune to the venom of coral snakes is still unknown.

What Do the Babies Look Like? Baby common kingsnakes generally look like miniature adults, although hatchlings of all the eastern subspecies may have visible yellow or white bands.

Distribution and Habitat Common kingsnakes occur throughout all or part of 19 eastern states, from the southern reaches of Illinois, Indiana, Ohio, and western West Virginia south to the Florida Keys and east to New Jersey and Delaware. The species has been reported from Pennsylvania (Lancaster and Chester Counties) but has not been documented since the 1950s. Carl Ernst of George Mason University considers the snake's existence in the state "probable or questionable." Kingsnakes are found in a wide variety of habitat types, including hardwoods, sandhills, grassy meadows, pine forests, savannas, and flatwoods; they also persist in suburban and agricultural areas. Although characteristically terrestrial, they often inhabit grassy shorelines of wetlands, including swamps, streams, and sometimes coastal salt marshes.

Behavior and Activity Kingsnakes are active primarily in the daytime. Although early evening activity has been observed during summer, nocturnal movement is relatively uncommon. Seasonal activity is similar to that of other eastern snakes in that they become inactive in late fall, hibernate through the winter, and become active again in early spring. In the New Jersey Pine Barrens kingsnakes have been observed hibernating in moving water in holes along the edges of streams. They will bask in the sun alongside burrows or on low vegetation during cool weather and may be found in or beneath rotting logs, rocks, trash, and other debris at all times of the year.

Food and Feeding Common kingsnakes are strong constrictors well known for eating other snakes, including venomous species—hence their name. They are immune to the venom of pit vipers and thus readily consume rattlesnakes, copperheads, and cottonmouths in addition to other snakes their size or smaller, and they have been known to consume snakes that are actually longer than they are. Common kingsnakes also eat rodents, lizards, birds and their eggs, and even small turtles. They often eat the eggs of other reptiles and are well known for burrowing through loose soil to excavate turtle nests. They have been documented to eat aquatic snakes that they capture in the water, suggesting that they may forage within the wetland itself. When eating small or defenseless prey, they often forgo constriction and simply swallow the prey alive. Common kingsnakes find their prey primarily through smell, although vision may be important as well.

Reproduction Common kingsnakes mate in the spring. During courtship and mating the male may bite and hold on to the female's neck. Approximately 10 eggs (range, 3–24) are laid in June or July within rotting logs, stumps, or similar moist, protected spots. The eggs, which tend to stick together in a clump, hatch 2–2.5 months later. Female kingsnakes sometimes produce clutches of eggs fertilized by multiple males.

Predators and Defense Kingsnakes face the same array of natural predators as most other snakes, but because of their relatively large size, adults need fear

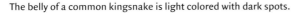

The belly of a common kingsnake is light colored with dark spots.

only the largest predators. When threatened in the wild, kingsnakes may vibrate the tail, assume a striking position, and release musk. They often bite if picked up or harassed. However, these defensive behaviors usually soon disappear and they can be easily handled.

Conservation Common kingsnakes are not protected by the federal government. The Outer Banks kingsnake, recognized by some herpetologists as a separate subspecies, has been listed as a species of special concern by the state of North Carolina. Kingsnakes are reported to be declining over much of their range. Although no single cause for reduced population sizes has been identified, many are killed on roads. Because kingsnakes are easily recognized and have a reputation for eating venomous snakes, people who ordinarily kill snakes on sight may spare them. Extensive collecting for the pet trade has been suggested as the cause of a possible decline of the Apalachicola kingsnake in western Florida.

What's in a Name? Carl Linnaeus named a specimen from "Carolina" *Coluber getulus* in 1766. He also referred to Mark Catesby's 1743 description and painting of the "chain snake" in *Natural History of Carolina, Florida, and the Bahama Islands*. Linnaeus's reason for choosing the name *getulus* is uncertain, but a suggestion that it refers to the geometric shapes characteristic of artwork by people of northern Africa known as Getulians seems plausible considering his description of the snake's appearance. The names for the various subspecies include one derived from Latin *niger*, meaning "black," for the black kingsnake (L. g. niger) and one from Greek *stiktos* and *kephos*, meaning respectively "spotted" and "head," for the Outer Banks kingsnake (L. g. sticticeps). Of the other three eastern subspecies, two were named in honor of famous herpetologists: the speckled kingsnake (L. g. holbrooki) for John Edwards Holbrook, author of *North American Herpetology*, and the Apalachicola kingsnake (L. g. goini) for Coleman J. Goin, the lead author of the textbook *Introduction to Herpetology*. The name L. g. meansi has also been proposed for the Apalachicola kingsnake in honor of Bruce Means. The Florida kingsnake (L. g. floridana) is named for the state of Florida.

Taxonomy of the common kingsnake is in a state of flux. Some authorities do not recognize some of the subspecies listed here, and others consider some of these subspecies to be full species. I see no compelling reason to alter the traditional taxonomy that common kingsnakes in the eastern United States belong to a single species with several subspecies. Colloquial names for the common kingsnake vary regionally because of the many color patterns the species exhibits, but the name chain kingsnake is frequently used for the eastern kingsnake (L. g. getula) and dates back to at least 1743.

PINE SNAKE *Pituophis melanoleucus*

SCALES
Keeled

ANAL PLATE
Single

BODY SHAPE
Heavy bodied; narrow, pointed head

BODY PATTERN AND COLOR
Light colored above with dark blotches; white belly

DISTINCTIVE CHARACTERS
Large scale covering a somewhat pointed nose

SIZE
baby
typical
maximum
0'   4'   8'

Description With the exception of the subspecies known as the black pine snake (*P. m. lodingi*), in which adults are typically solid black above and below, pine snakes characteristically have dark blotches on a lighter background. The light and dark markings are more contrasting in the northern pine snake subspecies (*P. m. melanoleucus*) than in the Florida pine snake (*P. m. mugitus*), in which the dark markings may be faded out and less visible, especially on the front part of the body and in more mature individuals. The pattern varies among regions and among individuals within regions, being lighter and more muted in some areas and more darkly contrasted in others.

top Northern pine snakes are light colored above with dark blotches and have a white belly.

What Do the Babies Look Like? Baby pine snakes generally look like the adults, but black pine snake babies are lighter colored than the adults and show signs of dark blotches instead of being solid black. Pine snake babies are longer and stouter than the adults of many species of eastern snakes.

Distribution and Habitat Pine snakes have been reported from 13 eastern states but are patchily distributed over much of their range, with the most extensive occurrence being throughout most of South Carolina and all but the southern tip of Florida. Only scattered localities are known in Tennessee, Kentucky, North Carolina, and Virginia. According to Tom Pauley of Marshall University, only a single confirmed record exists for the species in West Virginia (Monroe County). The most northern and eastern populations are in the southern New Jersey Pine Barrens. The black pine snake occupies a limited range from southern Alabama and Mississippi to Louisiana. Pine snakes generally live in areas with well-drained, sandy soil where they can easily burrow in search of prey. They are found in longleaf pine sandhill areas, pine barrens, scrub oak, dry rocky areas at high elevations in the Appalachian Mountains, and abandoned agricultural fields.

The Florida pine snake is lighter in color and the dark markings may be faded and less visible.

Adult black pine snakes are mostly black above and below. They were listed as threatened in 2015 under the U.S. Endangered Species Act.

PINE SNAKE *Pituophis melanoleucus*

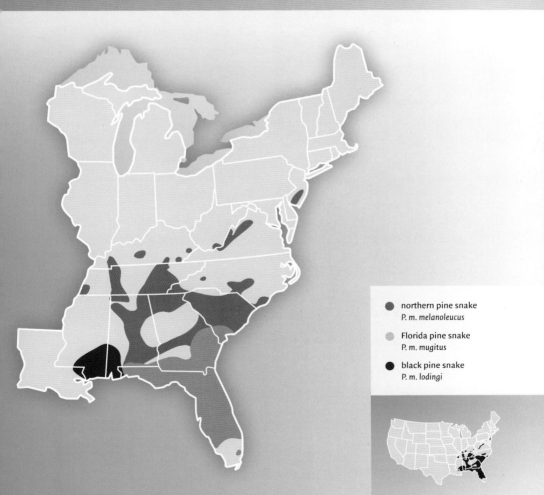

- ● northern pine snake
  P. m. melanoleucus
- ● Florida pine snake
  P. m. mugitus
- ● black pine snake
  P. m. lodingi

Pine snakes excavate a burrow in sandy soil to lay their eggs deep underground.

A rare observation of mating behavior by a pair of northern pine snakes.

Behavior and Activity Pine snakes are rather inactive from late fall to early spring throughout their geographic range, and hibernate in most regions. During the active season they spend most of their time underground in burrows that they dig themselves or that were dug by small mammals or gopher tortoises, or in tunnels left by decayed roots. They are most active on the surface in the spring or summer but may be found aboveground during warm winter months in the southern parts of their range. Surface activity occurs almost exclusively during daylight hours. They are capable of climbing into bushes and small trees but do so infrequently. Pine snakes often have very large home ranges that can cover several hundred acres. Adult males sometimes move long distances overland in search of mates, and females may travel as much as a mile to deposit eggs, often returning to the same nesting area.

Food and Feeding Pine snakes are extraordinarily powerful constrictors that consume a variety of mammals, including rabbits, squirrels, rats, mice, moles, and pocket gophers. They forage actively both above and below ground and have been observed climbing trees in search of nesting birds and their eggs. Small prey items (e.g., juvenile rodents) are eaten alive, without constriction. Juvenile pine snakes occasionally eat lizards.

Reproduction Pine snakes mate in the spring, and males presumably find females by following their pheromone trails. When a male finds a female, he rubs alongside her and may bite her neck to prevent escape. Male combat has been reported. During May, June, or July, females dig long burrows in loose soil and lay eggs at the end. Egg clutches range in size from 3 to 24 eggs (usually 9 or 10). Females nest communally in some portions of the geographic range, so that several clutches of eggs may be found in the same burrow. The eggs are among the largest of North American snakes, and the unusually large babies hatch about 2.5–3 months after the eggs are laid.

Predators and Defense Because pine snakes are so large, even as hatchlings, the primary predators are large birds of prey and larger mammals such as coyotes. Pine snakes encountered in the wild will often part their lips and hiss loudly, sometimes from several feet away, making a sound similar to that of a large rattlesnake. A small cartilaginous structure in front of the glottis helps the snake make the loud hissing noise. When picked up by a person they may coil tightly around their captor's arm, and they sometimes bite.

Conservation Habitat destruction resulting from urbanization is probably the single greatest threat to pine snakes. Because of their large size, generally slow crawling speed, and tendency to be active during peak daytime traffic periods, pine snakes are particularly vulnerable to road mortality. Hence, rural roads are among the greatest hazards they face in many parts of their geographic range. Of more than 100 pine snakes recorded in a South Carolina study, nearly 85 percent were snakes observed crossing or already dead on highways. Pine snakes are also victims of intentional human persecution because of their superficial resemblance to rattlesnakes. The black pine snake (*P. m. lodingi*) was listed as threatened in 2015 under the U.S. Endangered Species Act.

What's in a Name? This species was described as *Coluber melanoleucus* in 1803 by François-Marie Daudin, who gave it the French common name *la couleuvre noire et blanc* (the black and white grass-snake). The scientific name is derived from the Greek words *melaina*, meaning "black," and *leukos*, meaning "white," and from *pitys*, meaning "pine," and *ophis*, meaning "snake." Thomas Barbour described the Florida pine snake as a subspecies (*P. m. mugitus*) in 1921. The word *mugitus* means "bellowing" in Latin, referring to the loud hissing made by all pine snakes. The black pine snake (*P. m. lodingi*) was named by Frank Blanchard in 1924 in honor of Henry Peder Löding, a noted Alabama herpetologist.

An open mouth and loud hissing sound are part of the typical behavioral display by a threatened pine snake.

LOUISIANA PINE SNAKE *Pituophis ruthveni*

SCALES
Keeled

ANAL PLATE
Single

BODY SHAPE
Heavy bodied;
narrow, pointed
head

**BODY PATTERN
AND COLOR**
Light colored above
with dark blotches;
white belly

**DISTINCTIVE
CHARACTERS**
Large scale covering
a somewhat pointed
nose

SIZE

Description The Louisiana pine snake resembles its eastern counterpart (P. m. *melanoleucus*) in having a noticeably pointed head and a large scale covering its nose. The basic body color grades from pale yellow-brown on the head to darker yellow toward the tail. Large, dark brown to black, squarish blotches are present on the back beginning in the neck region, with dark spotting on the sides. The head and neck may have a network of brown reticulations. The dark markings on the sides are more interconnected on the front half of the body than on the posterior, where they become distinct from adjacent ones. The belly is white, but some individuals have brown markings.

top The Louisiana pine snake is similar in appearance and behavior to its eastern counterpart but is geographically isolated.

LOUISIANA PINE SNAKE *Pituophis ruthveni*

What Do the Babies Look Like? Newborn Louisiana pine snakes generally look like the adults. As with the pine snake, the babies are exceptionally large compared with other native snakes.

Distribution and Habitat The Louisiana pine snake is geographically isolated from its closest relatives, the pine snake (*P. m. melanoleucus*) and bullsnake (*P. catenifer*), being restricted to a limited area in west-central Louisiana and east-central Texas. In the early 2000s the species was confirmed to be present in only 4 Louisiana parishes and 5 Texas counties, and assumed to be extinct from most of its original range, which encompassed approximately 24 Louisiana parishes and Texas counties. This species is associated with sandy soils and the longleaf pine savannah habitat that formerly covered much of the region.

Behavior and Activity Louisiana pine snakes hibernate in pocket gopher burrows during the winter months and emerge in March. They remain active through the spring and early summer, becoming less so in late summer. A decided burst of activity is also apparent in the fall prior to cold weather. They spend most of their time underground and appear to be highly dependent on mammal burrows, predominantly those of pocket gophers, where they retreat on a daily basis. When encountered aboveground, Louisiana pine snakes immediately attempt to retreat to a burrow for safety. They move over a relatively small area daily during most of the year, but the total home range can be almost 200 acres for males. Surface activity occurs almost exclusively during daylight hours with very limited movement during the night.

Food and Feeding Louisiana pine snakes are powerful constrictors. Pocket gophers are their primary prey, but they probably eat other mammals as well, such as rabbits, squirrels, rats, and mice.

Scientists consider the Louisiana pine snake to be close to extinction and have proposed it as a candidate species for listing under the Endangered Species Act.

Reproduction Louisiana pine snakes mate in the spring and lay eggs in early summer about a month after mating. The eggs hatch about 2 months later. This species has very small clutches of three to five large eggs that are about 5 inches long and 2 inches wide. The average clutch number and egg size are biologically significant as the smallest and largest respectively of any species of native U.S. snake.

> **DID YOU KNOW?**
>
> *Sometimes snakes congregate around a particularly suitable habitat, but unless they are mating, they do so because of the environmental conditions, not because other snakes are there.*

Predators and Defense Because of their large body size and subterranean habits, adult Louisiana pine snakes have few natural predators. Their most likely natural enemies are large hawks and carnivorous mammals such as coyotes that might encounter them on the surface. Like other pine snakes, this snake hisses loudly when threatened.

Conservation The Louisiana pine snake may be close to extinction and has been proposed as a candidate species for listing under the Endangered Species Act. The list of threats they face is extensive, but habitat destruction as a result of commercial logging of natural longleaf pine forests is considered the primary cause of their perilous decline over most of their former geographic range. Road mortality is an additional and continuing hazard for these long, slow-moving snakes that are most likely to cross highways during daylight hours when traffic levels are highest.

What's in a Name? This species was described by Olive Griffith Stull in 1929 as a new subspecies of P. melanoleucus based on two specimens that she indicated differed in color and had a "larger number of dorsal spots" than other subspecies. The subspecies was named in honor of Alexander Ruthven, a noted herpetologist who became president of the University of Michigan. Herpetologists disagree about the taxonomy of the Louisiana pine snake; some consider it a separate species and others view it as a subspecies of the pine snake (P. melanoleucus) or of the bullsnake (P. catenifer).

BULLSNAKE *Pituophis catenifer*

SCALES
Keeled

ANAL PLATE
Single

BODY SHAPE
Moderately stout; narrow, pointed head

BODY PATTERN AND COLOR
Yellowish above with dark blotches; white belly with dark markings

DISTINCTIVE CHARACTERS
Large scale covering a somewhat pointed nose

SIZE
baby
typical
maximum
0' 4' 8'

Description The single subspecies of bullsnake (P. c. sayi) found in the eastern states has the basic body plan of the pine snakes. The body color is generally yellowish with large, dark brown or black blotches on the neck and back. The blotches on some individuals have a reddish tinge. The dark blotches form a banded pattern on the tail. The belly is yellow with dark spots.

What Do the Babies Look Like? Newborn bullsnakes have a more muted coloration but otherwise have the same pattern of blotches as the adults.

Distribution and Habitat Bullsnakes have an extensive western and southwestern geographic range, and isolated relict populations are found in three

top A bullsnake from Sauk County, Wisconsin

Juvenile bullsnakes have the same pattern of blotches on the body as adults.

BULLSNAKE *Pituophis catenifer*

Bullsnakes may assume a defensive pose and hiss when threatened.

eastern states: western and a small area in eastern Illinois, Wisconsin, and western Indiana. Bullsnakes generally live in what were formerly sand prairies or other areas with well-drained, sandy soils where they can use their pointed snout to burrow in search of prey.

Behavior and Activity Bullsnakes emerge from hibernation by March or April and remain active aboveground and in underground burrows until September or October. They are noted for hibernating communally and may travel overland long distances in the spring and fall between warm-weather foraging sites and winter refuges. They are most active on the surface in the spring but more likely to retreat underground during the summer. Bullsnakes forage aboveground mostly during the day but are occasionally active in early evening, especially during warm weather.

Food and Feeding Bullsnakes feed primarily on small mammals such as mice, rats, moles, ground squirrels, tree squirrels, and rabbits. They are powerful constrictors and have also been observed to press smaller prey tightly against the ground or other surface without completely wrapping around it. They typically kill mammals before consuming them but may eat very small, defenseless prey alive. Bullsnakes also eat turtle eggs, lizards, snakes, and birds and their eggs.

Reproduction Bullsnakes mate in the spring upon emerging from hibernation, mostly in April and May. The females usually lay an average of 11 large eggs (range, 2–24) in June, although later nesting has been reported. The eggs are typically about 3 inches long. The babies are longer and heavier than most other native snakes at birth.

Predators and Defense Known or suspected predators of bullsnakes are numerous and include large birds of prey such as hawks, great horned owls, and golden eagles. Larger mammal predators include raccoons, badgers, and coy-

Bullsnakes occur in the remnants of sand prairie habitats in three eastern states—Wisconsin, Illinois, and Indiana.

otes, and skunks and foxes probably prey on the eggs and young. Bullsnakes, like their close relatives the pine snakes, hiss loudly when threatened. Some herpetologists believe the hissing mimics the sound made by rattlesnakes when threatened.

Conservation Habitat destruction and roadway mortality are probably the greatest threats in the eastern states. People sometimes mistake bullsnakes for venomous snakes and kill them because their defensive display appears so threatening, but knowledgeable farmers and ranchers like to have bullsnakes around their barns because they help to control rodent populations.

What's in a Name? H. M. D. Blaionville described and named this species as *Coluber catenifer* in 1835. The Latin word *catena* means "chain mail armor," which is descriptive of the body pattern. In 1837 H. Schlegel recognized the eastern subspecies, *P. c. sayi*, and named it in honor of Thomas Say, a charter member of the Philadelphia Academy of Natural Sciences and the great-nephew of the early American naturalist William Bartram, author of the classic *Bartram's Travels*.

RATSNAKE *Pantherophis obsoletus*

SCALES
Weakly keeled

ANAL PLATE
Usually divided

BODY SHAPE
Slender to moderately stout and shaped in cross section like a loaf of bread

BODY PATTERN AND COLOR
Solid black or with blotches or stripes, depending on locality

DISTINCTIVE CHARACTERS
Relatively flat belly

SIZE
baby
typical
maximum
0'   4'   8'

Description Ratsnakes range from slender to relatively stout; some are as slim as racers, and others are heavier bodied. Presumably the individual's girth reflects its recent diet. Few North American snakes show as much regional variation in body pattern and color as the ratsnakes do, from the almost solid-colored black ratsnake (P. o. obsoletus) found in northern areas to the yellow ratsnake (P. o. quadrivittatus), which is yellow with black stripes, in Florida. The Everglades ratsnake (P. o. rossalleni) ranges from bright orange to yellowish orange with indistinct striping. The widespread gray ratsnake (P. o. spiloides) characteristically has a very light to dark gray body with large, darker gray blotches down the back and smaller ones along the sides. The subspecies prevalent throughout

top A typical black ratsnake from Clarke County, Georgia, showing the mostly solid black body and white chin

most of southern Louisiana, the Texas ratsnake (*P. o. lindheimerii*), resembles the gray ratsnake in pattern, but most specimens are more brown. The flat belly is generally lighter in color than the back, although black ratsnakes often have white under the front part of the body. Several zones of intergradation are apparent where black, Texas, gray, and yellow ratsnakes have contiguous ranges.

What Do the Babies Look Like? Newborn ratsnakes from most areas look like miniature versions of the gray or Texas ratsnake, with dark gray or brown blotches on a lighter gray body. They assume their regional adult coloration and pattern as they grow older. Baby Everglades ratsnakes may be yellowish with less distinctive blotches.

left The Everglades ratsnake is one of the most colorful eastern snakes.
middle The yellow ratsnake has a yellow chin and body with dark brown or black stripes the length of the body.   *right* A black ratsnake from Warren County, Ohio, with lighter coloration

The gray ratsnake has a gray body with large, darker gray blotches down the back and smaller ones along the sides.

The chin and belly of a black ratsnake in the neck region are generally white with mottling on the rest of the belly.

RATSNAKE *Pantherophis obsoletus*

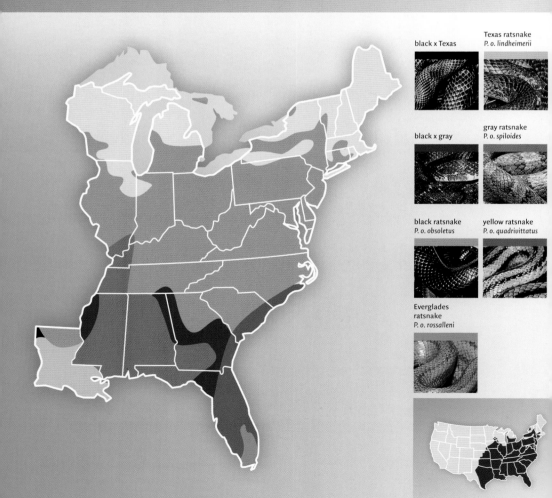

black x Texas

Texas ratsnake
P. o. lindheimerii

black x gray

gray ratsnake
P. o. spiloides

black ratsnake
P. o. obsoletus

yellow ratsnake
P. o. quadrivittatus

Everglades ratsnake
P. o. rossalleni

Greenish ratsnakes are intergrades between black or gray ratsnakes and yellow ratsnakes in coastal Georgia and the Carolinas. They are dull green or olive with black or brown stripes.

Although the adults may be almost black, brown, or yellow, the babies of most subspecies of ratsnakes look similar to boldly patterned gray ratsnakes.

Distribution and Habitat Ratsnakes are native throughout all or part of every eastern state except New Hampshire and Maine. They live in a wide variety of wooded and forested habitats but can occur in more open habitats as well. They are common in many suburbs, especially if the area still has many large trees in which they can climb, and are usually the most abundant or only large snake in many communities. They also frequent abandoned buildings and barns where they can hunt rodents and birds. They are often found near the edges of forests and also inhabit swamp forests and river floodplains.

Behavior and Activity Ratsnakes can be found during most times of the year, but during extremely cold or prolonged winters they may hibernate for 2–4 months in rotten stumps, hollow trees, or abandoned houses. During spring they are often active on the ground, in trees, and in flooded swamps. They are generally most active in daytime, especially when climbing in trees and bushes in the spring, but often prowl on warm nights both on the surface and in trees and shrubs in search of roosting birds or nests. Ratsnakes in southern parts of

A yellow ratsnake from Polk County, Florida, has the typical markings of this species in southern Florida.

A gray ratsnake from Albany, Georgia

the range hibernate in stump holes or root channels and beneath logs, rocks, or outbuildings. Communal hibernation occurs in northern and mountainous parts of the range, often with other species of snakes, including timber rattlesnakes.

Food and Feeding Ratsnakes are constrictors and active hunters that use both vision and smell to find their prey. Adults prey primarily on small mammals, birds, and bird eggs, which they typically swallow whole. Young ratsnakes feed mostly on treefrogs, small lizards, and baby rodents.

Reproduction Ratsnakes are egg-layers that generally mate during April, May, and June, and possibly in the fall in some areas. When one male encounters another during the breeding season, they sometimes engage in a wrestling match, presumably in competition for a female. During courtship and mating, the male frequently bites the female's neck. The clutch size ranges from 4 to 44 eggs, but the usual number is about 15. Eggs are laid in stump holes, tree holes, or other dark, moist situations, and several females may nest together. Ratsnakes have been known to return to the same area to nest year after year. The eggs hatch in about 2 months.

Predators and Defense A variety of snakes (e.g., copperheads, kingsnakes, racers, and indigo snakes), birds of prey (both hawks and owls), and mammals (raccoons, weasels, and coyotes) have been reported to prey on ratsnakes. Native predators that occupy the same habitats eat the eggs and juveniles. Wild ratsnakes encountered crawling on the ground commonly exhibit a kinking behavior in which the stretched-out snake makes a series of irregular kinks along the length of its body and then remains motionless; this configuration apparently serves as a form of camouflage and may help break up the linear pattern of the snake. It may also make the snake resemble a dead branch lying on the ground. Ratsnakes may bite when picked up, but if handled gently, many do not. They also may wrap around the captor's arm or the head of a predator and exude a very unpleasant musk from their cloacal (anal) glands.

Conservation Domestic cats and dogs kill many small ratsnakes in suburban areas. Many adult ratsnakes are victims of human persecution, and people sometimes kill the boldly marked juveniles because they mistake them for copperheads. Ratsnakes are frequent inhabitants of both suburban and farm communities and will not hesitate to eat chicken and duck eggs or small chickens and ducklings. Ratsnakes caught in the act are often killed. Ratsnakes climbing trees in pursuit of bird eggs are likely to attract the attention of birds, particularly blue jays, whose loud mobbing cries are likely to attract people. Along with bird eggs, however, ratsnakes eat enormous numbers of rats and mice wherever they live, and people who cannot appreciate their beauty or respect their right to share the land should value them as rodent-control agents. Some enlightened poultry owners view the loss of a few eggs as a small price to pay for a highly efficient

A threatened ratsnake displaying kinking behavior

rat and mouse patrol. One of the major threats to ratsnakes—people's irrational fear of them—can be lessened by educating the public to understand that recognizing another species as a predator is insufficient justification to kill it.

What's in a Name? Thomas Say described this species in 1823 and named it *Coluber obsoletus*. The Latin word *obsoleta* means "worn-out," and Say's reason for using it to describe the ratsnake is ambiguous. The Latin word *panther* means "cat" or "panther," but Leopold Fitzinger's reasons for using *Pantherophis* as the generic name in 1843 are unknown. Holbrook described the yellow ratsnake (*P. o. quadrivittatus*) as a full species in 1836, the name being based on Latin for *quad*, meaning "four," and *vittata*, meaning "wearing," in reference to the four distinct longitudinal stripes on the body. The Texas ratsnake (*P. o. lindheimerii*), described in 1853, was named after Ferdinand Jacob Lindheimer, who captured the first specimen near San Antonio. In 1854 Duméril, Bibron, and Duméril described the gray ratsnake (*P. o. spiloides*), whose name derives from Greek *spilos*, meaning "spots," referring to the blotched body pattern. The Everglades ratsnake (*P. o. rossalleni*), described by Wilfred T. Neill in 1949, was named in honor of E. Ross Allen of the Reptile Institute in Silver Springs, Florida.

The taxonomy of ratsnakes has been under intensive review by herpetologists, most of whom support separation of the New World ratsnakes (genus *Pantherophis*) from the Old World ratsnakes (genus *Elaphe*), although some continue to use the generic name *Elaphe* for New World ratsnakes. Additionally, certain genetic studies have been interpreted to suggest that rather than being partitioned into subspecies, the current species *P. obsoletus* actually comprises three distinct species: the eastern ratsnake (*P. alleghaniensis*), the western ratsnake (*P. obsoletus*), and the gray ratsnake (*P. spiloides*), with no recognized subspecies. However, the proposed species designations within this widespread group of snakes are not necessarily consistent with color and body pattern variation by which most people identify snakes. Also, these putative species have broad regions of geographic contact where they interbreed, as is characteristic of subspecies. Consequently, I use the traditional subspecies here because in most cases ratsnakes from particular geographic regions are easily identifiable based on color and pattern.

Known as black snakes in northern parts of their range, ratsnakes have traditionally been called chicken snakes in rural areas throughout much of the South since at least the 1700s. Oak snake and goose snake are used in some localities. Subspecies crossbreeding between *P. o. obsoletus* and *P. o. quadrivittatus* in coastal areas of Georgia and the Carolinas produces intergrades that are dull green or olive with black or brown stripes and are sometimes called greenish ratsnakes.

CORN SNAKE *Pantherophis guttatus*

SCALES
Weakly keeled

ANAL PLATE
Divided

BODY SHAPE
Moderately proportioned, with a flat belly and shaped in cross section like a loaf of bread

BODY PATTERN AND COLOR
Body reddish, yellowish, orange, or brown, with large blotches on the back and smaller ones on the sides; belly a black-and-white checkerboard with varying amounts of red; black stripes on the underside of the tail

DISTINCTIVE CHARACTERS
Lines on head meet between the eyes, forming a forward-pointing triangle that resembles a spear point

SIZE
baby
typical
maximum
0'   4'   8'

Description Some corn snakes can be among the most colorful snakes in the eastern United States with their bright red or orange body with darker blotches down the back and sides; others are duller variations of orange and brown; but all have two black-bordered stripes that extend forward across the head and connect to form a point between the eyes. A stripe behind the eye extends onto the scales of the neck. In western Louisiana, the corn snake may hybridize with the Great Plains ratsnake (*Pantherophis emoryi*), resulting in individuals that are drabber than the eastern forms. Coastal corn snakes are often more

top Corn snakes can be very colorful with a bright red or orange body with darker blotches bordered by black down the back and sides.

Baby corn snakes, as seen in these hatchlings from a clutch in New Jersey, have less orange or red than larger juveniles and adults.

reddish than inland forms, and their markings are often prominently outlined with black.

What Do the Babies Look Like? Baby corn snakes have bolder patterns than the adults, but usually less orange or red.

Distribution and Habitat Corn snakes occur throughout all or a large part of every coastal state except Maryland from New Jersey and Delaware to Louisiana. They are also found in eastern Tennessee and in isolated populations in Kentucky. Records from the 1970s exist from extreme northeast West Virginia (Morgan County), but recent documentation is limited. Corn snakes occupy a variety of habitats, including pine stands, abandoned agricultural fields, and hardwood forests. They are sometimes found near human habitations, including in wooded suburbs, and can be common under debris left around vacant houses and barns.

Behavior and Activity Corn snakes hibernate for several months in the northern portions of their range but may be active virtually year-round in the southern and warm coastal regions. When first becoming active in the spring they confine their aboveground movements to the daylight hours, but they become more nocturnal during the summer. In the fall they are again more likely to be active during daylight hours. During the winter they may surface to bask in the sun on warm days. Corn snakes frequently climb and are often found beneath the bark of dead trees.

Food and Feeding The diet consists mostly of small mammals but also includes birds and their eggs, lizards, other snakes, and frogs. Juveniles eat mostly small treefrogs and lizards. Corn snakes kill large prey by constriction and swallow smaller animals alive. They are active foragers and may hunt underground for rodents, moles, and voles, which they locate primarily by odor.

CORN SNAKE *Pantherophis guttatus*

left Corn snakes have a black-bordered stripe on each side of the head extending forward to connect and form a point between the eyes. Sometimes an aries symbol is present on the top of the head.   *right* The contrast of black and white markings on the belly is characteristic of corn snakes.

Reproduction Like other members of the genus *Pantherophis*, corn snakes are egg-layers. Mating begins in the early spring. While courting, the male moves slowly along the female's back, stimulating her with undulations of his body. Eggs are laid in June and July over most of the range, but a little earlier in southern Florida. Clutches may number 3–40 eggs, but the typical clutch size is about 14. Hatching takes place about 2 months after the eggs are laid.

A corn snake in the process of eating a vole after constricting it

Predators and Defense Natural predators include larger mammals, birds of prey (both hawks and owls), and kingsnakes. When threatened, corn snakes generally try to escape, but if unable to do so may vibrate the tail and bite. Most herpetologists do not consider their musk to be as disagreeable as that of their close relative, the ratsnake, nor are they prone to release such copious amounts. Corn snakes occasionally bite when first picked up but usually settle down quickly in captivity and make excellent pets.

Corn snakes from the Florida Keys are usually rosy in color.

Conservation Aside from natural predators and habitat destruction by humans, highway mortality is one of the greatest threats to corn snakes. Occasionally people kill corn snakes because they mistake them for copperheads. Corn snakes are sometimes collected for the pet trade, but the widespread and highly successful propagation of corn snakes among hobbyists has greatly diminished collecting of wild individuals.

What's in a Name? This species was formally described by Carl Linnaeus in 1766, presumably based on a specimen from Charleston, South Carolina. The Latin word *gutta* means "spot," although corn snakes are not really spotted. Linnaeus, however, referred to page 60 in Mark Catesby's 1743 book and the account and painting of "the bead snake" (which looks like no snake found in the eastern United States), which is described as having "large spots of a bright red colour; between which, at regular distances, are yellow spots." Presumably Linnaeus accepted Catesby's description that corn snakes are spotted (and therefore used the name *guttatus*). Ironically, another painting and account on page 55 in the same book clearly describe a corn snake, of which Catesby says, "It is all over beautifully marked with red and white, which seems to have given it the name of corn-snake; there being maize or Indian corn much resembling this in colour." Some herpetologists consider corn snakes in western Louisiana to be a separate species, Slowinski's corn snake (*P. slowinskii*), based on genetic studies. This species was named in honor of Joe Slowinski from the California Academy of Sciences, who died in Burma (Myanmar) from the bite of a banded krait on September 11, 2001. Corn snakes are sometimes called red ratsnakes.

Corn snakes will often assume a defensive pose, ready to strike if cornered.

GREAT PLAINS RATSNAKE *Pantherophis emoryi*

SCALES
Weakly keeled

ANAL PLATE
Divided

BODY SHAPE
Moderately proportioned, with a relatively flat belly and shaped in cross section like a loaf of bread

BODY PATTERN AND COLOR
Body gray with large, dull brown or olive blotches on the back and smaller ones on the sides; belly a black-and-white checkerboard; dark longitudinal stripes on the underside of the tail

DISTINCTIVE CHARACTERS
Dark lines on head converge to a point between eyes

SIZE
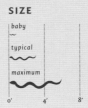

Description Great Plains ratsnakes have body patterns similar to those of corn snakes except that the background is generally gray, olive, or brown; and instead of having reddish blotches they have dark brown blotches down the back and sides. The borders around the blotches are dark but not usually black. Light and dark stripes extend forward across the head and converge to a point on the top of the head.

What Do the Babies Look Like? Baby Great Plains ratsnakes have colors and patterns similar to the adults.

top Great Plains ratsnakes have body patterns similar to those of corn snakes except that the background and blotches are brown or gray, rather than red or orange.

GREAT PLAINS RATSNAKE *Pantherophis emoryi*

Similar to corn snakes, light and dark stripes extend forward across the head of a Great Plains ratsnake and usually converge to a point.

Distribution and Habitat Great Plains ratsnakes occur primarily west of the Mississippi River with the exception of a small region in western Illinois where they have been documented from three counties (Jersey, Monroe, and Randolph) that border the river. They are found in a wide variety of habitats including rocky areas, woodlands, prairie grasslands, and farmlands.

Behavior and Activity Great Plains ratsnakes hibernate during the late fall and winter months in Illinois and other northern parts of their range. They are generally secretive during the day and hide beneath rocks or other ground cover. During warmer parts of the year they come aboveground and are active mostly at night.

Food and Feeding Great Plains ratsnakes are constrictors that primarily eat mammals but will also eat birds, bird eggs, snakes, lizards, and frogs. Rodents, shrews, bats, small opossums, and small rabbits have all been documented as prey. They actively seek out rodents in burrows.

Reproduction Great Plains ratsnakes mate from March through May and lay an average of 10 (range, 3–24) eggs from May through July. The seasonal timing is presumably earlier in warmer parts of the range. The eggs hatch about 2 months after they are laid.

Predators and Defense Large raptors (both hawks and owls), mammals, and other snakes are confirmed predators. The first response of a Great Plains ratsnake when threatened is to attempt to escape, but if cornered or restrained it will vibrate its tail and may bite and release musk.

Baby Great Plains ratsnakes hatching from their eggs. Note that the babies look similar to the adults.

A Great Plains ratsnake from Illinois. The species occurs west of the Mississippi River except for a few counties that border the river in western Illinois.

Conservation Like most other snakes, Great Plains ratsnakes suffer as a result of development and other human activities. Two of the greatest human-associated threats to them are road mortality and habitat destruction. Wildfires and prescribed burns are another source of mortality.

What's in a Name? This species was described by S. F. Baird and C. Girard in 1853 based on a specimen from Howard Springs, Texas, and named *Scotophis emoryi* in honor of William Hensley Emory, chief army surveyor of the United States and Mexican Boundary Survey in 1847–48. Baird and Girard noted the close similarity of the new species to the corn snake (*P. guttatus*), and some authorities have considered it to be a subspecies. Some herpetologists view Slowinski's corn snake (*P. slowinskii*; see page 203) as a hybrid or intergrade of *P. emoryi* and *P. guttatus*. This species is sometimes referred to as Emory's ratsnake.

FOX SNAKE Pantherophis vulpinus

SCALES
Weakly keeled

ANAL PLATE
Usually divided

BODY SHAPE
Moderately stout

BODY PATTERN AND COLOR
Brownish yellow with dark blotches on back and sides

DISTINCTIVE CHARACTERS
Head brown, coppery red, or orange

SIZE

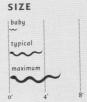

Description Fox snakes are moderately stout, although some individuals may be more slender depending on recent feeding opportunities. The basic body color is light brown, buff, or yellowish with distinct squarish blotches down the center of the back from head to tail with alternating smaller blotches on the sides. The belly is yellow with black markings. The eastern fox snake (P. v. gloydi) characteristically has fewer but larger blotches on the body. The top of the head is variable in color, with brown, reddish, and orange all being represented in some areas of Michigan. The eastern form is generally more reddish compared with the brownish head of the western fox snake (P. v. vulpinus), although the latter sometimes has a bright coppery head.

top The body color of fox snakes is light brown, buff, or yellowish with distinct squarish blotches down the center of the back from head to tail that alternate with smaller blotches on the sides.

What Do the Babies Look Like? Baby fox snakes look like pale versions of the adults but with dark markings on the head.

Distribution and Habitat Fox snakes have a more limited geographic range than other ratsnakes (genus *Pantherophis*) found in the eastern United States. Both of the two subspecies (western fox snake, *P. v. vulpinus*; eastern fox snake, *P. v. gloydi*) are found in the eastern states. The eastern fox snake is restricted to areas around the margins of Lake Huron and Lake Erie in southeastern Michigan and north-central Ohio. The range of the western fox snake includes northern Indiana and Illinois, most of Wisconsin, and most of the Upper Peninsula of

FOX SNAKE *Pantherophis vulpinus*

- western fox snake *P. v. vulpinus*
- eastern fox snake *P. v. gloydi*

Michigan. The two subspecies are separated from each other by more than 100 miles. Fox snakes are found in a diversity of habitats including open prairies, woods, and farmlands. The eastern fox snake is particularly noted for living in and around marsh habitats.

Behavior and Activity Fox snakes generally are active from April to October and hibernate the rest of the year. They are most likely to be encountered aboveground in May or June and are most active during the daytime, although surface activity has been reported at night during summer rains. Fox snakes will climb trees and bushes as well as into the rafters of barns. Communal hibernation has been reported, and several observations have been made of fox snakes hibernating underwater in old wells or cisterns. Animal burrows, muskrat or beaver lodges, and areas beneath the foundations of barns, houses, or other buildings can also serve as hibernation sites.

Food and Feeding Like the other ratsnakes, fox snakes are strong constrictors and active hunters. They have been documented to eat virtually all species of rodents within their geographic range as well as other small mammals, including baby rabbits, and birds and their eggs. Juvenile fox snakes eat frogs, lizards, and invertebrates (insects and earthworms).

Reproduction Fox snakes begin courtship and mating activities in May and June, which include male-male combat displays. Eggs are laid between June and August, mostly in July. They lay an average of about 14 eggs (range, 7–29). Communal nesting by several females has been reported. Hatching begins in August and may extend into October for late nests. Eggs have been found in

Eastern fox snakes like this one from Ohio often have a reddish head.

A western fox snake from Wisconsin

Two male fox snakes engage in combat during the mating season near Detroit, Michigan.

stump holes, rotting logs, and sawdust piles, and are probably also laid in other underground locations such as root tunnels and mammal burrows.

Predators and Defense Raptors, including large hawks and bald eagles, are known predators, and carnivorous mammals are likely predators as well. Mice will eat the eggs of fox snakes. Fox snakes generally do not bite when picked up, but will vibrate their tail and spray musk when threatened. Juvenile fox snakes are reported to respond more defensively by striking and biting.

Conservation People who kill snakes on sight and highway mortality are continuing threats to this species. Ironically, fox snakes are sometimes mistaken for venomous copperheads and killed in parts of their eastern geographic ranges where copperheads do not occur. The loss of marsh habitats to commercial real estate and agricultural development has unquestionably taken its toll on fox snake populations, especially eastern fox snakes.

What's in a Name? This species was described and named *Scotophis vulpinus* by S. F. Baird and C. Girard in 1853 based on a specimen provided to the Smithsonian by Dr. Philo Romayne Hoy from Racine, Wisconsin. A second specimen from Grosse Ile, Michigan, provided by Rev. Charles Fox, was also used to describe the species and is the origin of the scientific name (*vulpes* means "fox" in Latin) and the common name. Some herpetologists have erroneously indicated that the common name refers to the pungent foxlike smell of this snake when captured. The eastern fox snake (*P. v. gloydi*) was named in honor of herpetologist Howard Kay Gloyd, who was director of the Chicago Academy of Sciences. Some authorities consider the two subspecies of fox snakes to be separate species.

RACER Coluber constrictor

SCALES
Smooth

ANAL PLATE
Divided

BODY SHAPE
Slender and streamlined

BODY PATTERN AND COLOR
Typically solid black, gray, or brownish; sometimes with light spots

SIZE
baby
typical
maximum
0'   4'   8'

Description Nine subspecies of racers are found in the eastern United States (five in Louisiana alone). Most are solid colored above and below. Three subspecies are black above and black or bluish below: the northern black racer (*C. c. constrictor*) and southern black racer (*C. c. priapus*) have white chins, and the brown-chinned racer (*C. c. helvigularis*) has tan to brown lips. The Everglades racer (*C. c. paludicola*), blackmask racer (*C. c. latrunculus*), eastern yellowbelly racer (*C. c. flaviventris*), and blue racer (*C. c. foxii*) are various shades of gray, green, or brown above and light yellowish to bluish below. The buttermilk racer

top Most racers, such as this northern black racer, are solid colored above and below.

RACER Coluber constrictor

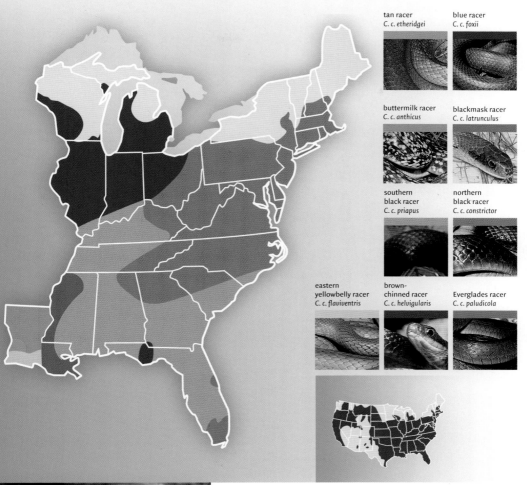

tan racer
C. c. etheridgei

blue racer
C. c. foxii

buttermilk racer
C. c. anthicus

blackmask racer
C. c. latrunculus

southern
black racer
C. c. priapus

northern
black racer
C. c. constrictor

eastern
yellowbelly racer
C. c. flaviventris

brown-
chinned racer
C. c. helvigularis

Everglades racer
C. c. paludicola

The dark marking behind the eye is visible on this blackmask racer from Louisiana.

above A blue racer from Indiana

right Juvenile racers differ from the adults in not being solid in color but in having dark blotches along the back and sides.

(*C. c. anthicus*) and tan racer (*C. c. etheridgei*) differ from the solid-colored racers in having light spotting on a darker body. Intergradation of subspecies may occur along zones of contact. All racers are slender and streamlined for speed.

What Do the Babies Look Like? Baby racers have proportionally larger eyes and differ from the solid-colored adults in having a light-colored body with dark blotches that may appear reddish along the back and sides.

Distribution and Habitat Racers occur in part of every eastern state, and statewide in all but six of them. They are absent from the northern parts of Wisconsin, Michigan, New York, Vermont, New Hampshire, and Maine. They occur in a wide variety of habitats but prefer open areas such as old agricultural fields and large grasslands, forests with sparse undergrowth, brushy areas, and the edges of marshes and swamps.

Behavior and Activity Racers hibernate for at least a few weeks each winter in most of their range north of Florida, and considerably longer at northern latitudes and higher elevations. They become active each year in early-to-mid spring. Racers seek refuge during the winter, as well as at other times of the year, beneath leaves and other ground litter, under logs, in stumps, or in burrows and tunnels in the soil. In more northern and colder regions they will seek out deeper, more sheltered hibernacula. Racers are active exclusively during the daytime, and an individual may travel a long distance during a single day. Racers frequently climb into trees or bushes.

Food and Feeding Racers probably have a more diverse diet than any other North American snake. It includes small rodents, lizards, other snakes (including smaller racers), birds, frogs, insects, and even small turtles. Racers will eat

The buttermilk racer is not solid-colored like most other racers but instead has light spotting on a darker body.

An Everglades racer from Everglades National Park, Florida

Brown-chinned racers have brownish lips. This one is from the panhandle of Florida.

the eggs of other reptiles and birds, and will even eat some venomous snakes such as copperheads and pigmy rattlesnakes. Cannibalism has been noted as well; a newly hatched racer was even observed to eat the next one that hatched. They are visually oriented, active hunters that often search for prey with the head held high like a periscope. When prey is spotted, they chase it down, seize it, and swallow it alive. Despite their scientific name, racers are not constrictors. Large or potentially dangerous prey may be chewed on until subdued and easier to swallow.

> **DID YOU KNOW?**
>
> Many nonvenomous snakes such as racers, kingsnakes, and ratsnakes vibrate their tails when frightened. If the tail is vibrated in dry leaves, it may produce a sound much like a rattlesnake's rattle.

Reproduction Racers mate over a several-month period that varies with the geographic area in which they live. In more northern latitudes they typically mate between April and July, but in Florida they may mate as early as February. They lay 4–36 eggs (usually 10–13) during the summer. As with many other species, larger females usually lay more eggs than smaller ones. The oblong eggs are readily identifiable because they are covered by tiny granules that look like salt. They are laid in moist, protected areas such as under fallen bark or large rocks, or buried a few inches underground. They hatch in about 2–2.5 months.

Predators and Defense Natural predators include birds of prey such as hawks, and larger mammals such as raccoons and foxes. When threatened, racers first use their speed to try to escape, often retreating into a bush or tree where they stop and remain motionless. They may also disappear under a rock or into an animal burrow. If escape is not possible, racers will not hesitate to defend themselves. A threatened racer generally vibrates its tail and will strike repeatedly at its tormentor. Racers are notorious biters when captured, and because of their quickness usually get in several good bites—often accompanied by intense chewing—before they can be restrained.

Conservation Habitat destruction and urbanization are the primary threats to racers. Racers are fast-moving snakes that do not seem to be intimidated by open areas such as fields, grasslands, and highways. Their inclination to cross roads gets many of them killed by automobiles.

What's in a Name? Carl Linnaeus named this species in 1758 based on a preserved specimen from Canada. The genus name is derived from the Latin word *coluber*, meaning "serpent," to which Linnaeus assigned many of the snakes he first described. The origins of the names of the nine eastern subspecies are varied. Two were named for individuals: the blue racer (*C. c. foxii*) in 1853 for Rev. Charles Fox, who provided a specimen from Michigan for the original description; and tan racer (*C. c. etheridgei*) in 1970 for Richard E. Etheridge, who

The eastern yellowbelly racer is found in southwestern Louisiana.

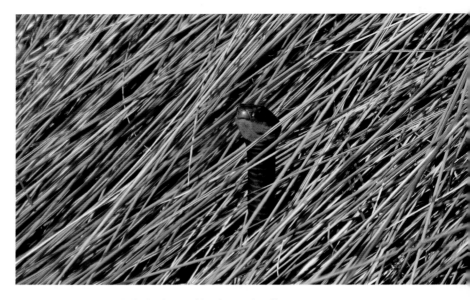

A racer uses periscoping behavior in searching for prey in tall grass.

collected the first specimen of the subspecies. Four of the subspecies names refer to the color pattern of the snake: the eastern yellowbelly racer (*C. c. flaviventris*), described in 1823, from Latin *flavi*, meaning "yellow," and *ventris*, meaning "belly"; buttermilk racer (*C. c. anthicus*), described in 1862, from Greek *anthikos*, meaning "white speckled"; brown-chinned racer (*C. c. helvigularis*), described in 1955, from Latin *helvus*, meaning "pale yellow," and *gula*, meaning "throat"; and blackmask racer (*C. c. latrunculus*), described in 1970, from Latin *latrunculus*, meaning "robber," referring to the black facemask. The southern black racer (*C. c. priapus*), described in 1939, was named after Priapus, the god of fertility, in reference to the structure of the hemipenes, which differentiates this subspecies from the others. The name of the Everglades racer (*C. c. paludicola*), described in 1955, is derived from Latin *palus*, meaning "swamp."

Although the original name, *Coluber constrictor*, is still used today, racers are not really constrictors, a trait that would not have been identifiable in the preserved animal available to Linnaeus. Racers are often called black runners or blue runners in the rural South.

COACHWHIP *Masticophis flagellum*

SCALES
Smooth

ANAL PLATE
Divided

BODY SHAPE
Long and slender

BODY PATTERN AND COLOR
Solid black on head and anterior back and belly, changing to tan on posterior portion of body

DISTINCTIVE CHARACTERS
Tail scales resemble a braided whip

SIZE
baby
typical
maximum
0'   4'   8'

Description Adult coachwhips are the only eastern snakes that are solid black anteriorly and completely tan on the posterior portion of the body. The proportion of black to tan varies considerably over the range; those from parts of northern Florida and southern Georgia, for instance, are mostly tan. They have large eyes, a relatively large head, and a tail that resembles a braided whip because of the way the scales are patterned. Of the six subspecies, only one, the eastern coachwhip (M. f. flagellum), occurs in the eastern United States.

top Coachwhips are the only eastern snakes that are typically solid black on the front half of the body and completely tan on the posterior portion. The name refers to the braided whip appearance of its scales from midbody to tail.

Coachwhip babies are not black on the front of the body but are light tan with irregular brownish bands along the back and sides that fade posteriorly.

What Do the Babies Look Like? Like racers, coachwhip babies do not look at all like the adults. They are light tan with irregular brownish bands along the back and sides that fade posteriorly.

Distribution and Habitat Coachwhips have been reported from nine eastern states, including all coastal states from North Carolina to Louisiana, as well as small portions of southwestern Tennessee and southwestern Illinois. According to Chris Phillips of the Illinois Natural History Survey, the species is probably extinct in Illinois because no individuals have been seen since the 1970s. John MacGregor of the Kentucky Department of Fish and Wildlife Resources has indicated that coachwhips are not native to Kentucky and that past records were based on escaped or released captive specimens from roadside reptile displays. Coachwhips inhabit open areas such as old fields, longleaf pine forests, palmetto flatwoods, and sandy scrub oak habitats, generally, but not always, in areas with sandy, dry soil.

Behavior and Activity Like most other eastern snakes, coachwhips in southern Florida and warm coastal areas are active on some days every month, and those in colder areas hibernate during most of the winter. They generally retreat underground into animal burrows, old root tunnels, or beneath logs or vegetation during cold weather and at night. They are active aboveground only during daylight hours, often in the hottest seasons and at times of day when most other snakes are inactive. They prefer more open habitats and sometimes climb into shrubs and small trees. Coachwhips often travel long distances during the active season, and the home range of an individual may be extensive.

Food and Feeding Like their relatives the racers, coachwhips actively search for prey during the daytime. They often hunt in open areas, with the head held high in periscope fashion and moving from side to side as they look for the small

opposite Some coachwhips are solid in color as seen in this individual from southern Georgia.

COACHWHIP Masticophis flagellum

mammals (e.g., mice, voles, and rabbits), lizards, birds, and other snakes on which they feed. They frequently eat large grasshoppers and other invertebrate prey, and food records include at least four species of turtles. They sometimes climb high into trees or bushes in pursuit of prey. Coachwhips do not constrict their prey, but grab it and swallow it alive. If several small rodents can be captured at once, a coachwhip will hold one or more down with its body while it swallows another.

Reproduction Mating usually occurs in April or May, shortly after hibernation. Courtship has not been documented in the wild, but there is evidence that males fight for females by engaging in a wrestling match in which each male tries to force the other's head to the ground. About 11 eggs (range, 4–24) having a granular appearance like those of racers are laid in June and July. Hatching takes place in the late summer or early fall.

Predators and Defense Natural predators of adult coachwhips are limited to larger animals such as coyotes and large hawks. Younger individuals may fall prey to other snake-eating snakes, raccoons, and foxes. Like racers, coachwhips first use their speed to try to escape any would-be predator, sometimes even climbing into bushes or small trees. If escape is not possible, a threatened coachwhip will strike repeatedly to defend itself and will hold its ground with an open mouth, usually also vibrating its tail. Coachwhips usually try to bite when first captured, although some will not bite if handled gently. Their agility and speed often results in their captor being bitten at least once or twice.

Conservation Loss of habitat resulting from urbanization and development is likely the greatest threat to coachwhips. They are wide-ranging snakes and thus are likely to encounter roads. Although they are among the faster-moving snakes, vehicles are a major threat to those that live near busy highways.

What's in a Name? This species was first described in 1743 by Mark Catesby, who referred to it as the "coach-whip," but did not receive a formal scientific name until George Shaw described and named it in 1802. Shaw is perhaps better known for his original descriptions of two other species, the duckbill platypus and the kiwi. The scientific name reflects the snake's appearance, with *mastix* in Greek and *flagellum* in Latin both meaning "whip." Coachwhips are sometimes called whipsnakes. Some herpetologists place the species in the genus *Coluber*.

EASTERN INDIGO SNAKE *Drymarchon couperi*

SCALES
Mostly smooth, although mature males have keeled scales down the middle of the back

ANAL PLATE
Single

BODY SHAPE
Moderately slender to robust

BODY PATTERN AND COLOR
Solid blue-black on back and undersides

DISTINCTIVE CHARACTERS
Chin and lips reddish, brown, or black

SIZE
baby
typical
maximum
0'   4'   8'

Description Adult indigo snakes are solid black, but their iridescent scales make them look gunmetal blue in the sunlight. The reddish or orange to brown coloration on the chin may extend onto the lips, face, neck, and anterior part of the belly, especially on juveniles and some adult males. They are the longest and largest native snakes inhabiting the United States.

What Do the Babies Look Like? Indigo snake babies resemble adults, but more of the chin and the area behind the head may be reddish. The spaces between

top Adult indigo snakes are solid black. The iridescent scales can make them look gunmetal blue in the sunlight.

Hatchling indigo snakes are not shiny black like the adults but may have bluish and white crossbands. Indigo snake babies are huge compared with those of most other eastern snakes.

EASTERN INDIGO SNAKE *Drymarchon couperi*

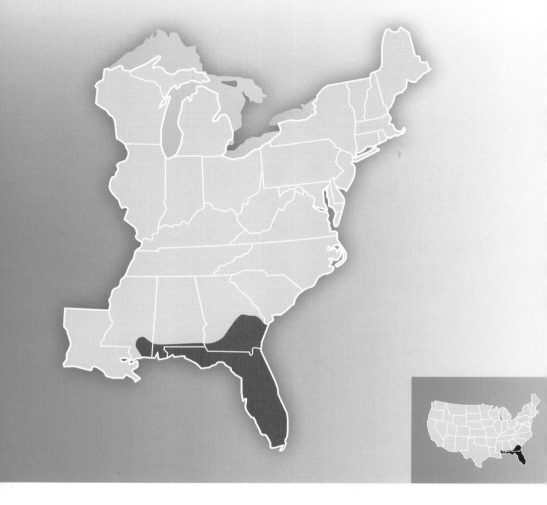

the scales on the back are cream colored or light blue, and some hatchlings have faint, creamy blue crossbanding or blotching. Compared with most other southeastern snakes, newly hatched indigo snake babies are huge, ranging from 16 inches to more than 2 feet long.

Distribution and Habitat The historical geographic range included all of Florida and southeastern Georgia and extreme southern Alabama into Mississippi. Today, natural populations in the Southeast are known only from Georgia and Florida. Indigo snakes prefer dry, open habitats with sandy, well-drained soils but occur in a wide variety of habitat types in central and southern Florida. They can be found in longleaf pine flatwoods and sandhills, turkey oak barrens, cabbage palm–live oak hammocks, palmetto prairies, and, in the Everglades region, tropical hardwood hammocks and mangrove communities. During warmer months they are often found at the margins of wetlands, where they patrol for prey.

Behavior and Activity Because of their strictly southern geographic range, indigo snakes may be dormant for only a few days or weeks during cold winters or not at all. In fact, a sunny day during winter is usually the best time to find indigo snakes as they bask in the sun to warm up. Favored retreats are stump holes and the burrows of gopher tortoises, armadillos, and small mammals. They are active aboveground only during daylight hours and have been known to travel up to 4 miles between their summer foraging areas and winter retreats.

The color of the chin and lips of an indigo snake varies regionally and can be reddish, brown, or black.

Even though indigo snakes can grow to more than 8 feet in length, they are generally docile when handled.

Indigo snakes are often associated with gopher tortoise burrows.

Food and Feeding Indigo snakes feed on a wide variety of prey. They are not constrictors but instead use their size and powerful jaws to overpower their prey. They frequently feed on other snakes, which they grab by the head and pin down with their heavy body, including venomous snakes such as rattlesnakes, copperheads, and cottonmouths. Their diet includes lizards, birds and their eggs, rodents and other small mammals, frogs, toads, turtles and their eggs, and even small alligators. Cannibalism in the wild has been reported.

Reproduction Indigo snakes have been observed mating in the wild from late October to February. During courtship the male rubs his chin along the female's back, holding her down until she raises her tail, indicating her readiness to mate. About 9 eggs (range, 4–17) are laid in May or June, probably in burrows made by pocket gophers or gopher tortoises or in stumps or other natural cavities in sandy soil. The extremely large eggs (up to 3 inches long) hatch about 3–3.5 months after they are laid.

> **DID YOU KNOW?**
>
> Among eastern snakes, several species live more than 20 years, including ratsnakes, indigo snakes, and copperheads. One timber rattlesnake lived more than 30 years in captivity, and another lived more than 40 in the wild.

Predators and Defense Because of their large size, indigo snakes that reach adulthood have few natural predators. Alligators, large raptors such as red-tailed hawks, and some medium- to large-sized mammals (bobcats, coyotes, and wild pigs) are known or suspected predators of adults. Juveniles are potential prey for a wider number of predators. When threatened, indigo snakes generally try to escape. They seldom bite when first captured or in captivity, but when they do, the strength of their powerful jaws makes it a memorable experience. Indigo snakes sometimes hiss and, unlike other southeastern snakes, often flatten the neck vertically when threatened.

Conservation The eastern indigo snake was officially designated as threatened throughout its range under the U.S. Endangered Species Act in 1978. At that time the species was known to occur in Florida and Georgia and had been reported historically in Alabama, Mississippi, and South Carolina, although the report from South Carolina (in the 1930s) has never been confirmed. Because the species is federally protected, indigo snake populations are likely no longer threatened by overcollection for the pet trade and also are less likely to be harmed by people who do not like snakes. Nonetheless, negative public attitudes and local ignorance about snakes (and about the legal implications of the Endangered Species Act) still result in unnecessary killing of these magnificent creatures in some areas. However, many private landowners in rural south Georgia and Florida revere and avidly protect indigo snakes on their lands.

An indigo snake from Telfair County, Georgia

Although not done as overtly as in the past, the illegal practice of pouring gasoline down gopher tortoise burrows during rattlesnake roundups continues in some areas and is potentially fatal to any indigo snake present. One of the greatest threats for a wide-ranging snake like the indigo is habitat destruction and degradation resulting from development, fire suppression, and elimination of suitable winter refuges.

Road mortality may be a significant threat in areas where traffic is high. Unfortunately, this species does not appear to have recovered since its federal listing as threatened, and populations continue to decline or disappear. A project to reintroduce indigo snake populations in southern Alabama is under way and if successful will aid in restoring the species in that part of its former geographic range.

The eastern indigo snake is protected under the U.S. Endangered Species Act.

What's in a Name? This species was originally named *Coluber couperi* by John Edwards Holbrook in 1842 in honor of J. Hamilton Couper, a Georgia plantation owner who was familiar with the biology of indigo snakes and noted their association with gopher tortoise burrows. The genus name is fitting in that *drymos* means "forest" and *arkhon* means "ruler." Herpetologists do not agree on the range-wide classification of indigo snakes. Some authorities consider the eastern indigo snake a subspecies of *Drymarchon corais*, which is found in south Texas southward through Central America into South America. Indigo snakes are called blue gophers or blue bullsnakes in some areas of the southeastern United States.

watersnakes

previous page Northern watersnake

BLACK SWAMP SNAKE *Seminatrix pygaea*

SCALES
Smooth

ANAL PLATE
Divided

BODY SHAPE
Moderately stout

BODY PATTERN AND COLOR
Solid black above with red belly

SIZE
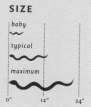

Description These small, moderately stout snakes have a shiny black back and a red or orange-red belly. The head is relatively small. Some females may be rather heavy bodied at certain times of the year, generally before they give birth. The smooth scales often have a light line running down the center, giving the impression that they are keeled. The three recognized subspecies (Carolina swamp snake, S. p. paludis; north Florida swamp snake, S. p. pygaea; and south Florida swamp snake, S. p. cyclas) are distinguished primarily by the number of belly scales and the amount of black on each belly scale.

What Do the Babies Look Like? Baby black swamp snakes are identical in appearance to the adults.

top The black swamp snake is the only eastern species that is shiny black above with a solid red belly and smooth scales.

Distribution and Habitat Black swamp snakes are known primarily from lower Coastal Plain counties from North Carolina to Alabama, most of Florida, and from a few localities in the upper Coastal Plain of South Carolina. They are usually associated with still, shallow waters of heavily vegetated wetlands and generally are not found in or around the moving waters of streams or rivers.

> **DID YOU KNOW?**
>
> The primary diet of some snakes is other snakes.

Behavior and Activity Black swamp snakes are characteristically most active in early-to-late spring but can be found in any month of the year if temperatures are warm enough. In colder regions they hibernate in the roots of aquatic vegetation, in muskrat lodges, and beneath vegetation or ground litter along the shore. Black swamp snakes are possibly the most aquatic snakes in the eastern United States and, unlike the larger watersnakes, seldom bask out of the water. Overland travel occurs during both day and night but may be restricted if conditions are too dry or windy because these snakes are more susceptible to desiccation than other snakes. Black swamp snakes found on land are generally close to a wetland edge and usually associated with dense aquatic vegetation. They survive droughts by burrowing into the substrate. Their populations suffer considerably less during extreme droughts than those of other aquatic snakes, and they appear to be particularly well adapted to seasonally dry isolated wetlands.

Although very secretive and seldom seen, black swamp snakes sometimes occur in large numbers in natural wetlands.

Food and Feeding Black swamp snakes eat a wide variety of animals found in and around wetland habitats, including leeches, earthworms, small fish, small salamanders and salamander larvae, frogs, and tadpoles. They presumably locate their prey primarily by scent and swallow it alive. In some locations aquatic amphibians, especially larval or fully aquatic salamanders, are preferred prey.

Reproduction Virtually nothing is known about the courtship and mating behavior. Mating probably occurs in late spring, and most females give birth to live young between July and September. Some evidence exists that females emerge onto dry land to give birth. Litter size usually ranges from 2 to 15

BLACK SWAMP SNAKE *Seminatrix pygaea*

- Carolina swamp snake
  S. p. paludis
- north Florida swamp snake
  S. p. pygaea
- south Florida swamp snake
  S. p. cyclas

(average, around 8) but may be more than 20. Large females have more babies than small females.

Predators and Defense Because of their small size, black swamp snakes often fall prey to wetland predators such as turtles, wading birds, raccoons, and even largemouth bass, as well as to kingsnakes. The bright red belly has been suggested to be a defense mechanism that distracts predatory birds such as herons or egrets that search for prey along the edges of wetlands. When a bird foraging among the debris along a muddy shoreline turns over the snake, the bright red belly catches the bird's attention; while the bird is looking for a

A black swamp snake from Craven County, North Carolina, near the northern edge of the geographic range of the species.

The bright red belly of the black swamp snake has been suggested to be a defense mechanism that distracts wading birds searching for prey around wetlands.

bright red prey item, the shiny black snake turns over and crawls to safety. The red belly has also been suggested to serve as a "warning" signal to predatory animals, although swamp snakes are not venomous. Black swamp snakes rarely bite when picked up.

Conservation Their dependence on wetland habitats makes black swamp snakes vulnerable to habitat loss, degradation, or isolation. The introduction of exotic fish is reported to have reduced some populations in Florida. Because black swamp snakes are rarely seen unless specifically searched for, their occurrence and subsequent decline due to habitat modifications in some areas may go undetected, and thus their conservation status remains murky.

What's in a Name? The first black swamp snake known to science was found in Volusia County, Florida, and originally named *Contia pygaea*. Edward Drinker Cope, who described it in 1871, said, "This interesting addition to our reptile fauna is quite unlike any species heretofore found in our territory." The scientific name is a derivative of the Latin *pygae*, meaning "rump." The reference to the rump is explained by a deformity or injury of the original specimen Cope used in the species description, saying, "about one-sixth of the length in front of and behind the vent [is] compressed." In 1895 Cope renamed the genus *Seminatrix*, a name derived from the Latin words *semi*, meaning "half," and *natrix*, meaning "watersnake," presumably because Cope considered the black swamp snake "halfway like" a watersnake. At that time it was still known only from Florida. The names of the subspecies described by Herndon Dowling in 1950 (S. p. cyclas and S. p. paludis) derive from the Latin words *cyclas*, "an outer garment with decorative border," and *palus*, meaning "swamp." Some authorities place the black swamp snake in the genus *Liodytes*.

QUEEN SNAKE *Regina septemvittata*

SCALES
Keeled

ANAL PLATE
Divided

BODY SHAPE
Slender to moderately robust

BODY PATTERN AND COLOR
Brown, grayish, or olive green above with yellowish stripes on the sides

SIZE

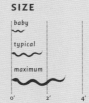

Description Queen snakes have a light brown, grayish, or olive green back with three darker, usually inconspicuous brown stripes running along the entire length. A yellow to cream-colored stripe runs down each side, and the belly is yellowish with four brown stripes running its length. Queen snakes are slender, although females may be more robust than males, especially just before they give birth. The head is not much wider than the neck.

What Do the Babies Look Like? Baby queen snakes look like the adults but may be lighter in color and have more distinct stripes.

Distribution and Habitat Queen snakes are found in 20 eastern states, but populations are often in widely separated areas. They are found in Piedmont

top Queen snake from Hall County, Georgia

Queen snakes frequently bask on tree limbs overhanging the water.

QUEEN SNAKE *Regina septemvittata*

and mountainous regions of West Virginia, Virginia, the Carolinas, Georgia, the eastern three-fourths of Tennessee, and central Kentucky. Their range extends throughout Alabama and into Coastal Plain streams in several areas including the panhandle of Florida and southern Mississippi. They have a limited distribution in northeastern Illinois, parts of southern Michigan and Wisconsin, most of Ohio, and the eastern half of Indiana. The species is known from eastern and southeastern Pennsylvania and western New York but is absent from Louisiana, New Jersey, and the New England states. These are the quintessential water-snakes of cold, clear, rocky streams lined with shrubby vegetation where crayfish are abundant. They have also been found in a variety of other aquatic habitats, including small reservoirs in Ohio, along the Potomac River in Virginia, and in sandy-bottomed streams in South Carolina and Alabama.

Behavior and Activity Queen snakes become inactive for several weeks or months during the winter in most regions, retreating inside root masses, under vegetation and ground litter near the shore, or into burrows made by other animals next to streams. In the most northerly parts of their range they hibernate below the freeze line in rocky crevices, man-made dams and other structures, and crayfish burrows. Several individuals are sometimes found together in communal dens. During warmer months, queen snakes can often be found under partially submerged large rocks or logs. They commonly bask on the limbs of shrubs and small trees overhanging the water in spring and fall, and occasionally bask during the summer as well. They are active primarily during the day.

Food and Feeding Queen snakes, like their close relatives, Graham's crayfish snakes, feed almost exclusively on defenseless newly molted crayfish, which they locate by detecting a hormone released by molting crayfish. They avoid hard-shelled crayfish, which can defend themselves with large pincers. In addition to crayfish, queen snakes sometimes eat small fish. All prey animals are grabbed and swallowed alive, generally tail first.

Queen snakes are livebearers that give birth in late summer or early fall.

Reproduction Like other members of the genus *Regina*, queen snakes apparently mate in the spring, although at least one observation indicates that they may have a fall mating season as well. Queen snakes give birth to 5–39 (average, about 11) live young during summer or early fall.

The light-colored side stripe and brown stripes on the belly are key characters for identifying a queen snake.

Predators and Defense Natural predators of queen snakes are primarily those associated with stream habitats or their margins and include racers, kingsnakes, cottonmouths, large fish, hellbender salamanders, and, ironically, perhaps even large crayfish. Queen snakes writhe and release musk when captured but rarely bite.

Conservation Like the three other species of crayfish-eating snakes, queen snakes depend on the environmental health of their habitat for survival, apparently because of their dietary dependence on crayfish. Thus, stream pollution, siltation (primarily from agricultural runoff), stream channelization, and other forms of degradation that reduce or eliminate crayfish populations are the primary threats to populations. Activities that reduce the amount of shoreline and natural retreats used for shelter can also be detrimental.

What's in a Name? The only reference to the queen snake's locality in Thomas Say's original 1825 description was that it "inhabits Pennsylvania." The name *septemvittata* is derived from the Latin words *septem*, meaning "seven," and *vittatus*, meaning "striped," in reference to the seven dark and light stripes that run the length of the body on the back and belly.

GRAHAM'S CRAYFISH SNAKE *Regina grahamii*

SCALES
Keeled

ANAL PLATE
Divided

BODY SHAPE
Slender to moderately stout

BODY PATTERN AND COLOR
Brown with a broad yellowish stripe on each side

SIZE

Description Graham's crayfish snakes are brown or grayish brown with a broad yellowish stripe running the length of each side and, usually, a dark stripe running down the middle of the back. Males are generally slender, but some females can be quite stocky. The head is not much wider than the neck. The belly is solid yellow, sometimes with a central row of spots.

What Do the Babies Look Like? Baby Graham's crayfish snakes look like the adults but may be more boldly marked.

Distribution and Habitat This species has one of the patchiest geographic distributions of any eastern U.S. snake. It is known only from scattered localities in Mississippi, Louisiana, and Illinois, where it is associated with a variety of wetland types, including small lakes and ponds, swamps, bayous, and slow-moving streams where crayfish abound.

top Graham's crayfish snakes can be recognized by the broad yellowish stripe on each side of a brown body.

Behavior and Activity Graham's crayfish snakes hibernate in colder regions, commonly using crayfish burrows as winter refuges. In spring and fall they are most active during the day, but in summer they are occasionally found at night. In some areas they are most often seen from March or April to June and virtually disappear in the late summer and fall, presumably taking refuge in crayfish burrows. Individuals may bask during the day on tree limbs, bushes, and other vegetation along the edges of wetlands. They move overland between aquatic habitats and are sometimes found on land along shorelines.

Food and Feeding As the name implies, Graham's crayfish snakes eat primarily crayfish, specializing on newly molted individuals with soft shells. They occasionally eat small frogs and minnows as well. They do not kill their prey before eating it, but simply swallow it alive.

Reproduction Graham's crayfish snakes mate in the spring (April and May) and give birth to live young in the late summer or early fall. Little is known about their courtship, but on several occasions multiple males have been observed trying to mate with one female. Litter size ranges from 9 to 39, with an average of about 15. During gestation, females apparently can transfer nutrients to the developing offspring much as mammals do, in contrast to many other live-bearing snakes whose embryos acquire their nourishment only from an attached yolk sac.

Predators and Defense Natural predators associated with wetlands, such as wading birds, large fish, and snake-eating snakes, will eat this snake. Graham's crayfish snakes are generally mild-mannered when captured but may thrash about and usually release a foul-smelling musk from glands at the base of the tail.

A Graham's crayfish snake basking along the shore of Lake Decatur in Macon County, Illinois

Conservation Habitat destruction and alteration that changes the hydrology of the wetlands they require is likely the biggest threat to Graham's crayfish snakes. Their strict dependence on healthy crayfish populations makes them particularly vulnerable to forms of habitat degradation that reduce crayfish populations.

What's in a Name? The name *grahamii* used in the original description by S. F. Baird and C. Girard in 1853 refers to Col. James Duncan Graham of the U.S. Army, who provided the original specimens to the Smithsonian. The snakes were collected in Texas during the United States and Mexican Boundary Survey that began in 1848.

GRAHAM'S CRAYFISH SNAKE *Regina grahamii*

Because of their dependence on a diet of recently molted crayfish, Graham's crayfish snakes are vulnerable to habitat degradation that reduces the health and abundance of crayfish.

STRIPED CRAYFISH SNAKE *Regina alleni*

SCALES
Smooth

ANAL PLATE
Divided

BODY SHAPE
Relatively slender to moderately stout

BODY PATTERN AND COLOR
Alternating brown and yellowish stripes

SIZE
baby
typical
maximum
0" 12" 24"

Description Striped crayfish snakes generally have three brown stripes running the length of their yellowish brown back. The back often has an iridescent sheen, especially when wet. The belly is cream colored and sometimes has a row of spots running along the center. Striped crayfish snakes are somewhat slender to moderately stout (especially adult females), and the head is not much wider than the neck.

The striped crayfish snake has alternating brown and yellowish stripes that run the length of the body.

What Do the Babies Look Like? Juvenile striped crayfish snakes look like the adults.

top Adult striped crayfish snakes subsist primarily on a diet of crayfish.

Distribution and Habitat Striped crayfish snakes are restricted to Florida east of the central panhandle and southeastern Georgia. They are associated with swamps and open wetlands with heavy vegetation and are not usually found around the moving waters of streams or rivers.

Behavior and Activity Striped crayfish snakes are active throughout most of the year except for cold periods during the winter. During active periods they may hide in the roots of aquatic vegetation, including the introduced water hyacinth, or on land beneath logs or debris. They can be active in the water during the day or at night. They have been reported to bask in sunny spots to warm themselves.

STRIPED CRAYFISH SNAKE *Regina alleni*

An unusual reddish colored striped crayfish snake from Everglades National Park, Florida

They can often be found crossing roads that pass through wetland areas, especially during rainy periods.

Food and Feeding Like other members of the genus *Regina*, striped crayfish snakes specialize on crayfish. Their chisel-like teeth help them to hold on to the hard carapace of the crayfish, which they nearly always swallow tail first. There are some reports of them subduing crayfish by constriction before swallowing them. Juveniles eat grass shrimp and insect larvae, especially the larvae of dragonflies.

Reproduction Very little is known about their reproduction. They apparently mate in the spring, and young are born in the late summer or early fall. Litter sizes range from about 4 to 12, and larger females produce more babies than smaller females.

Predators and Defense Documented predators include great egrets, great blue herons, sandhill cranes, kingsnakes, cottonmouths, large aquatic salamanders (sirens and amphiumas), alligators, and river otters. When disturbed, striped crayfish snakes try to escape into the water. Individuals release musk from scent glands in the cloaca when captured and may writhe open mouthed in the captor's hand, but most do not bite.

Conservation Because of their specialized diet, striped crayfish snakes can disappear from an area if pollution or other habitat degradation eliminates their primary prey. Therefore, in addition to wetland destruction resulting from human development, one of the greatest threats to crayfish snakes is habitat alteration that reduces crayfish populations.

What's in a Name? S. W. Garman described this species in 1874 from a specimen captured near Jacksonville, Florida, by Joseph Asaph Allen, for whom it is named. Allen was an ornithologist who was the first president of the American Ornithologists' Union and the author of Allen's Rule, the biological principle that birds and mammals in cold climates have shorter and thicker extremities than those in warm climates. The word *regina* in Latin means "queen." Some authorities place the striped crayfish snake in the genus *Liodytes*.

GLOSSY CRAYFISH SNAKE *Regina rigida*

SCALES
Smooth

ANAL PLATE
Usually divided

BODY SHAPE
Stout

BODY PATTERN AND COLOR
Solid brown or gray above with whitish belly

SIZE
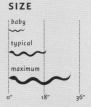

Description The body of glossy crayfish snakes is dark brown or olive green with two narrow, light stripes running down the center of the back. The back usually has an iridescent sheen if wet. The yellow or cream-colored belly has two rows of spots running down the center. The head is not much wider than the neck. The three subspecies (glossy crayfish snake, R. r. rigida; Gulf crayfish snake, R. r. sinicola; and Delta crayfish snake, R. r. deltae) are distinguished by slight variations in color pattern and scale numbers. Populations with many solid black individuals have been found in the Mississippi Delta and in Florida.

What Do the Babies Look Like? Juvenile glossy crayfish snakes look like the adults, although they may be more boldly marked and their bellies may be somewhat pink.

top Glossy crayfish snakes vary little in appearance throughout their range.

Distribution and Habitat Glossy crayfish snakes are restricted to the Coastal Plain of the southeastern United States and appear to have a spotty distribution, although that may be a result of inadequate documentation. They occur from Virginia into the southeastern halves of the Carolinas; in parts of central and northern Florida; in the southern halves of Georgia, Alabama, and Mississippi; and in most of Louisiana in swamps, floodplains, and open wetlands with heavy vegetation. They are not usually found around swift-moving streams or rivers.

Behavior and Activity Glossy crayfish snakes are active throughout most of the year except for cold periods during the winter, when they usually hibernate in crayfish burrows. They have been reported to be active at night and during the day during summer rains, although behavior of this species has rarely been documented. They can sometimes be found basking on tree limbs in river swamps on sunny days during late winter and early spring.

Food and Feeding Like other members of the genus *Regina*, glossy crayfish snakes specialize on crayfish. Their chisel-like teeth fit into the grooves on crayfish shells, giving the snake a secure hold. They may constrict their prey and nearly always swallow it tail first. Other reported prey items include small frogs, fish, and small rodents. Juveniles eat insect larvae, especially the larvae of dragonflies, and presumably juvenile crayfish.

Glossy crayfish snakes sometimes bask on cool sunny days in late winter or spring.

GLOSSY CRAYFISH SNAKE *Regina rigida*

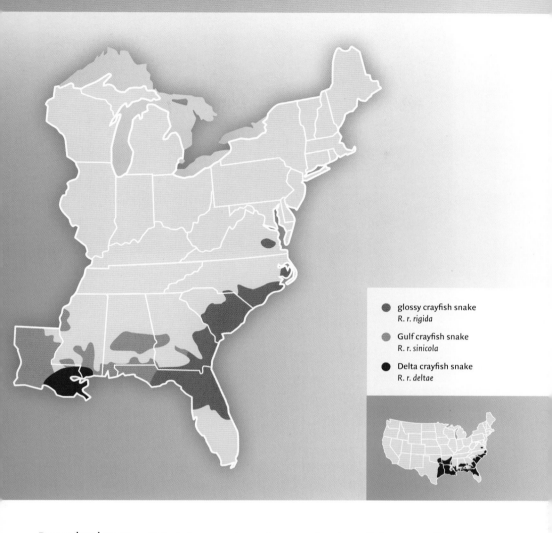

Reproduction Very little is known about the reproduction of glossy crayfish snakes. They apparently mate in the spring, and young are born in the late summer or early fall. Litters can range from about 6 to 19 young, and larger females probably produce more babies than smaller females do.

Predators and Defense Because these snakes are rarely seen, documentation of predators is difficult. They likely fall victim to any larger vertebrate that might prey on a snake of their size. Known predators include large aquatic salamanders (sirens and amphiumas), alligators, and kingsnakes. When disturbed,

Two rows of dark spots on the yellowish belly give the appearance of stripes. This glossy crayfish snake is from Aiken, South Carolina.

they try to escape into the water. If captured they thrash around, release musk from their cloacal glands, and will sometimes gape, but they rarely, if ever, bite.

Conservation Records of glossy crayfish snakes are too scarce to determine their status in most regions. They are particularly rare in North Carolina and Virginia, and because there are so few locations where populations are known to occur, they should be considered species of special concern. Destruction of wetlands and swampy areas has likely eliminated many undocumented populations. Pollution or other forms of habitat degradation that eliminate the crayfish that are their primary prey will result in the loss of this species as well.

What's in a Name? This species was the first of the crayfish snakes that Thomas Say described, in 1825, stating that it "inhabits the southern states" and "frequents the water." Although *rigidus* means "stiff" or "rough" in Latin, the connection of the name with this species, described on the basis of a single preserved specimen, is not obvious. The names used for the two subspecies, named in 1959, refer to their general geographic ranges. *Regina r. deltae* occurs in the Mississippi River delta. The name *R. r. sinicola*, the subspecies that inhabits the Gulf region, was derived from Latin *sinus*, meaning "gulf or bay," and *incola*, meaning "an inhabitant." Some authorities place the glossy crayfish snake in the genus *Liodytes*.

NORTHERN WATERSNAKE *Nerodia sipedon*

SCALES
Keeled

ANAL PLATE
Divided

BODY SHAPE
Heavy bodied

BODY PATTERN AND COLOR
Variable dark bands that change to dark blotches on the rear that alternate with smaller blotches on the sides

SIZE
baby
typical
maximum
0' 3' 6'

Description Both coloration and pattern are highly variable. Typically, northern watersnakes have bands of reddish brown and lighter brown on the front part of the body changing to alternating square blotches on the back and sides toward the rear of the body. Females reach larger body sizes than males, and older adults often become solid dark brown or even dull black. The belly is typically yellowish or cream colored with one or more half-moon spots on each belly scale. Four subspecies have been described. The northern watersnake (*N. s. sipedon*), which has 30 or more body bands or blotches, and the midland watersnake (*N. s. pleuralis*), which has fewer than 30, are both wide-ranging subspecies. The Carolina watersnake (*N. s. williamengelsi*) is typically much darker than the other subspecies. The Lake Erie watersnake (*N. s. insularum*) generally

top The dark bands on the front part of the body of northern watersnakes become dark blotches on the rear and alternate with smaller blotches on the sides.

Northern watersnakes frequently bask on rocks or logs.

NORTHERN WATERSNAKE *Nerodia sipedon*

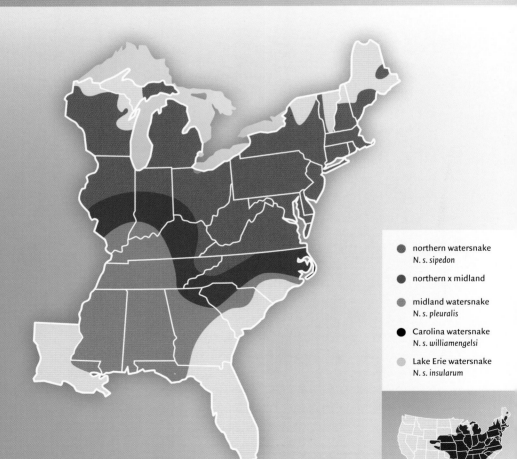

- northern watersnake
  N. s. sipedon
- northern x midland
- midland watersnake
  N. s. pleuralis
- Carolina watersnake
  N. s. williamengelsi
- Lake Erie watersnake
  N. s. insularum

Northern watersnakes are highly variable in body color and pattern throughout their range and even within the same population.

has a pale body with the characteristic banded pattern being vague but apparent in many individuals.

What Do the Babies Look Like? Babies have greater contrast between dark and light crossbands and blotches.

Distribution and Habitat Northern watersnakes are found in part of every eastern state and throughout the state in most of them. The northern watersnake subspecies (*N. s. sipedon*) ranges from eastern Tennessee and western North Carolina northward, and the midland watersnake is found from southern Illinois and Indiana south. The two subspecies have a broad region of intergradation from Illinois to North Carolina. The Carolina watersnake (*N. s. williamengelsi*) is found in saltwater and brackish habitats only on the Outer

A midland watersnake from Georgia

Banks of North Carolina and the adjacent mainland. The Lake Erie watersnake (*N. s. insularum*) occurs in northern Ohio on the Put-in-Bay Islands in Lake Erie. Northern watersnakes are likely to be found in any aquatic habitat within their geographic range, including suitable habitat on coastal barrier islands. Fish hatcheries, small streams, large rivers, swamps, marshes, Carolina bay wetlands, bayous, ponds, lakes, and reservoirs are home to this species. They are commonly found in backyard ponds in suburban areas, where they sometimes eat ornamental fish.

Behavior and Activity Northern watersnakes hibernate during the winter and during the warmer months retreat into a variety of hiding places such as root masses; burrows of other animals; overhanging shorelines; and nearshore vegetation, logs, and rocks. In northern regions individuals have been observed traveling overland to hibernation sites several hundred feet from water, and some in southern populations may do the same. During sunny days in the spring and fall, and cooler periods in the summer, northern watersnakes often bask on trees and shrubs along the shore, always with a ready escape route to the water. They may be active during the daytime in spring and fall but in southern areas are more likely to be active in the water at night during warm periods of the year. Northern watersnakes in some areas make frequent journeys overland between aquatic habitats.

Food and Feeding Northern watersnakes have been documented to eat a greater variety of fish and amphibians than any other North American watersnake; more than 80 fish species and 30 amphibians have been recorded as prey.

Although they eat mostly vertebrates, invertebrates such as insects, earthworms, and leeches are eaten, especially by juveniles, when the opportunity arises. In addition to being active foragers, northern watersnakes will wedge themselves among rocks and sway back and forth with the mouth open ready to catch unsuspecting fish; they sometimes corral fish in shallow water with a body coil. All prey items are swallowed alive, generally headfirst.

Reproduction Northern watersnakes mate during late April, May, and early June. It is not uncommon for many males to try simultaneously to mate with a single, larger female. Females sometimes mate with more than one male and bear litters with offspring produced by multiple fathers. Litter sizes range from 4 to nearly 100, but 20–30 is typical. The mother nourishes the developing em-

left Large adult northern watersnakes are often melanistic.   right The Carolina watersnake subspecies is typically much darker than the other subspecies.   bottom The northern subspecies of the northern watersnake from Michigan

bryos through a placenta-like structure, and birth takes place in late summer or early fall.

Predators and Defense Adults and juveniles have a wide variety of natural predators, including many species of birds, mammals, turtles, other snakes, and predatory fish. A threatened watersnake will try to escape, but if that is not possible it will spread its jaws slightly and flatten the front part of the body to make the head look bigger and will strike and bite to defend itself. If picked up, northern watersnakes characteristically writhe, bite, and release musk from their anal glands.

Conservation Northern watersnakes are common over most of their range because they can live in a variety of aquatic habitats. Unfortunately, they are frequent victims of anglers and wildlife agents who mistakenly believe that they prey on trout and other game fish at a level that affects fishing success. Many harmless northern watersnakes are mistaken for venomous cottonmouths and killed, even in areas hundreds of miles from where cottonmouths live. The Lake Erie watersnake (N. s. insularum) is considered a threatened species by the state of Ohio.

What's in a Name? Carl Linnaeus named this species *Coluber sipedon* in 1758. The origin of *sipedon* is uncertain. The subspecies N. s. *williamengelsi* was named in recognition of William L. Engels, who collected watersnakes on the North Carolina barrier islands before leaving for World War II. He returned to become a professor of zoology at the University of North Carolina. The midland watersnake subspecies (N. s. *pleuralis*) name is from the Greek *pleuralis*, meaning "the side," and refers to the banding pattern on the sides of the body. The Lake Erie watersnake (N. s. *insularum*) gets its name from the Latin *insula*, meaning "island," referring to its geographic location in the islands in Lake Erie. Some herpetologists do not accept N. s. *insularum* as a valid subspecies because individuals move extensively between the islands and the mainland and interbreed with N. s. *sipedon*.

In some areas of the South this species is traditionally known as the banded watersnake both because of its appearance and because the northern watersnake and banded watersnake were once considered the same species, the latter previously being known as *Nerodia sipedon fasciata*.

SOUTHERN BANDED WATERSNAKE *Nerodia fasciata*

SCALES
Keeled

ANAL PLATE
Divided

BODY SHAPE
Heavy bodied

BODY PATTERN AND COLOR
Dark crossbands on a light brown body

DISTINCTIVE CHARACTERS
Dark brown or black stripe behind eye and onto neck

SIZE
baby
typical
maximum
0'   3'   6'

Description Southern banded watersnakes are stout snakes that typically have irregular dark brown or reddish brown bands on a lighter background running the length of the body. Large adults sometimes become solid dark brown or dull black. The belly is yellowish, and there are red spots or wavy markings on each belly scale. All three subspecies (banded watersnake, N. f. fasciata; Florida watersnake, N. f. pictiventris; and broad-banded watersnake, N. f. confluens) occur in eastern states. The broad-banded watersnake has fewer (about 10–11) and much wider bands than the other two subspecies.

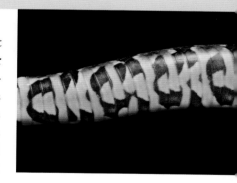

Banded watersnakes have a yellowish belly with red markings on each scale.

top Most southern banded watersnakes have brown or reddish cross bands from head to tail.

What Do the Babies Look Like? Babies have banding similar to that of the adults but with sharper contrast between dark bands and background coloration.

Distribution and Habitat Southern banded watersnakes occur in almost every aquatic habitat in the Coastal Plain of the Carolinas and Georgia, throughout Florida, in the southern portions of Alabama and Mississippi, throughout Louisiana, and up the Mississippi River drainage to extreme western localities in Tennessee, Kentucky, and Illinois. They can be found in small and large streams, rivers, swamps, marshes, bayous, and essentially all ponds, lakes, and reservoirs. They can be particularly abundant around ephemeral wetlands such as Carolina bays, where they feed on abundant amphibians.

Behavior and Activity Southern banded watersnakes are active year-round in the southern parts of their range and become inactive during cold periods in winter in the other areas. They hibernate in animal burrows near the water, in muskrat and beaver lodges, and under fallen logs or shoreline vegetation. They

Florida watersnakes can be highly variable in body color and pattern.

SOUTHERN BANDED WATERSNAKE *Nerodia fasciata*

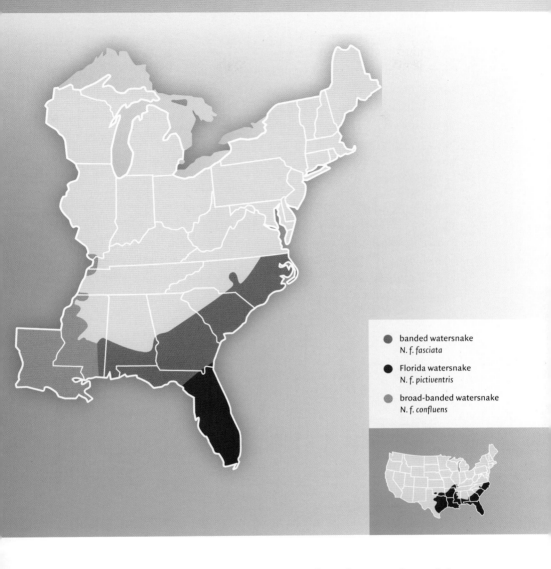

typically retreat to similar places during warmer weather. They are active at night in many areas during the summer but also may be active during the day. On cool, sunny days they often bask, sometimes on limbs overhanging the water.

Food and Feeding Southern banded watersnakes are active foragers that eat a variety of aquatic animals, mostly fish and amphibians. Jeff Camper reported one from South Carolina that had eaten a musk turtle. Most fish are swallowed

A southern banded watersnake in search of prey in a wetland

headfirst, and most frogs are swallowed rear first. Prey are not subdued but are swallowed whole while still alive.

Reproduction Like other North American watersnakes, southern banded watersnakes typically mate during the spring and give birth to live young from midsummer into fall. Litters range from 6 to as many as 80 young but typically number about 20–25. Multiple males sometimes congregate in shallow-water habitats during the early spring to mate with receptive females.

Predators and Defense Cottonmouths, great blue herons, and alligators are all known predators of southern banded watersnakes, as are numerous other predatory animals that live in and around aquatic ecosystems within their range. When threatened, a southern banded watersnake nearly always tries to escape first. If cornered, it will flatten its head and body, attempting to appear larger and more formidable. If captured, it will often twist around, bite, and release copious amounts of musk.

A large melanistic banded watersnake from coastal South Carolina

The broad-banded watersnake has fewer bands than the other two eastern subspecies.

Conservation People in many regions kill watersnakes on sight, and the southern banded watersnake is no exception. They are often mistaken for copperheads because of the banded color pattern or for cottonmouths because of their presence in and around water. Any human activities that result in wetland habitat destruction or degradation are threats to populations of southern banded watersnakes. An additional danger results from their tendency to cross highways that are adjacent to their wetland habitats.

What's in a Name? In 1766 Carl Linnaeus named this snake from "Carolina" *Coluber fasciatus*. The species name comes from the Latin word *fascia*, meaning "band," which refers to the banded body pattern. The subspecies name *pictiventris* refers to the patterned belly and is derived from the Latin words *pictus*, meaning "painted," and *ventris*, meaning "belly." The name *confluens* is derived from the Latin word *confluere*, meaning "to flow together," which refers to the broad, coalescing banding pattern characteristic of this subspecies.

SALT MARSH SNAKE *Nerodia clarkii*

SCALES
Keeled

ANAL PLATE
Divided

BODY SHAPE
Moderately stout to heavy bodied

BODY PATTERN AND COLOR
Extremely variable: banded, solid, or striped

SIZE
baby
typical
maximum
0' 2' 4'

Description Salt marsh snakes exhibit a bewildering variety of color patterns. The back and sides may be solid, striped, or blotched, and the belly may be solid or spotted. Body colors include brown, black, yellow, white, and red in various combinations above and below. Three subspecies are recognized. The Gulf salt marsh snake (*N. c. clarkii*) is the least variable in color pattern, usually having four longitudinal stripes on a lighter background. The mangrove salt marsh snake (*N. c. compressicauda*) may be solid, banded, blotched, or even somewhat striped. The Atlantic salt marsh snake (*N. c. taeniata*) is also highly variable but usually has stripes on the first one-third of its body. Intergrades occur in the areas where the ranges of the subspecies overlap, and this species may hybridize with southern banded watersnakes in some areas.

top Salt marsh snakes vary greatly in body color and pattern. The Gulf salt marsh snake, shown here, usually has light and dark stripes that run the length of the body.

Salt marsh snakes may hybridize with southern banded watersnakes in some areas, as seen in these individuals from Everglades National Park (*left and middle*) and Hickory Mound Wildlife Management Area in Florida (*right*).

SALT MARSH SNAKE *Nerodia clarkii*

- Atlantic salt marsh snake
  N. c. taeniata
- mangrove salt marsh snake
  N. c. compressicauda
- Atlantic x mangrove
- Gulf salt marsh snake
  N. c. clarkii
- Gulf x mangrove

> **DID YOU KNOW?**
>
> The eastern United States is home to three snakes (indigo snake, Atlantic salt marsh snake, black pine snake) that have been placed on the federal endangered species list.

What Do the Babies Look Like? Baby salt marsh snakes are as variable as the adults, but their stripes and spots are usually more distinctive. Mangrove salt marsh snakes can have striped, solid, and blotched babies in a single litter.

Distribution and Habitat The salt marsh snake is unique among North American snakes in being partial to brackish water and inhabiting the coastal margins in each of the Gulf Coast states. The only Atlantic Coast populations are in Florida. The species is commonly associated with salt marsh and mangrove swamps, but individuals will occasionally enter freshwater habitats adjacent to brackish water areas.

Behavior and Activity Salt marsh snakes are active year-round in southern Florida but are inactive for short intervals during winter along most of the Gulf Coast. They bask less frequently than the other large watersnakes and are usually inactive during the day, hiding in vegetation, in animal burrows (including those made by crabs), and under logs or marsh grass. They forage in the water, mainly at night, and their activity is greatly influenced by tidal cycles in some areas.

Food and Feeding Salt marsh snakes primarily eat small fish that live in brackish water, which they may lure by wiggling and curling the tongue in the water and at the water's surface. They occasionally eat crustaceans such as fiddler crabs.

Reproduction The reproductive biology is poorly known, although they apparently mate during the spring and give birth in late summer and early fall. Litter size ranges from 1 to 24 with an average of about 10.

Predators and Defense Natural predators of salt marsh snakes are those associated with brackish-water habitats and include marine fish, herons and egrets, hawks and other raptors, and large crabs. They rely on camouflage for protection and usually react more mildly to capture than the larger watersnakes,

Salt marsh snakes frequently bask in mangrove trees.

Mangrove salt marsh snake swimming through young mangroves in the Florida Keys

sometimes not even biting. Like other watersnakes, however, they usually release musk when captured.

Conservation In 1977 the subspecies known as the Atlantic salt marsh snake was officially designated as threatened throughout its entire range (Volusia County, Florida) under the federal Endangered Species Act. Deterioration of natural barriers that occurs when canals are constructed between brackish and freshwater habitats can result in interbreeding between salt marsh snakes and their close relatives, southern banded watersnakes, and result in loss of the population's genetic integrity. Most populations may be imperiled in the near future as a result of coastal development.

Mangrove salt marsh snakes from the Florida Keys are the most variable subspecies in color and pattern.

What's in a Name? S. F. Baird and C. Girard in 1853 named this species *Regina clarkii* based on specimens provided to the Smithsonian by Col. J. D. Graham that were collected by John H. Clark in Indianola, Texas. Robert Kennicott described the mangrove salt marsh snake (N. c. *compressicauda*) in 1860, referring to the body as "much compressed toward the tail." He combined the Latin words *com*, meaning "together," *pressus*, meaning "pressed," and *cauda*, meaning "tail," to compose the name. Despite his description, typical specimens do not have a compressed tail, so the individual used in the original description was apparently aberrant. The name of the Atlantic salt marsh snake (N. c. *taeniata*), described by Edward Drinker Cope in 1895, is derived from the Latin word *taeniatus*, meaning "striped," and refers to the striped pattern on the anterior of the body.

PLAIN-BELLIED WATERSNAKE *Nerodia erythrogaster*

SCALES
Keeled

ANAL PLATE
Divided

BODY SHAPE
Heavy bodied

BODY PATTERN AND COLOR
Usually a solid brown back and plain yellow to red belly

SIZE
baby
typical
maximum
0'   3'   6'

Description Plain-bellied watersnakes are stout snakes with a brown, gray, or greenish gray back and, as the common name implies, a plain, unmarked belly ranging in color from yellow to red. Four subspecies occur in the eastern United States: the red-bellied watersnake (*N. e. erythrogaster*) has a red to orange belly; the yellow-bellied watersnake (*N. e. flavigaster*) has a light orange to yellow belly; and the blotched watersnake (*N. e. transversa*) has a yellow belly and a faded pattern of alternating blotches on its back. The copperbelly watersnake (*N. e. neglecta*) typically has a black or dark brown body and a red belly that can range from scarlet to reddish orange. They may also have some black markings on the sides.

top Plain-bellied watersnakes are characterized by a plain, unmarked belly.

PLAIN-BELLIED WATERSNAKE *Nerodia erythrogaster*

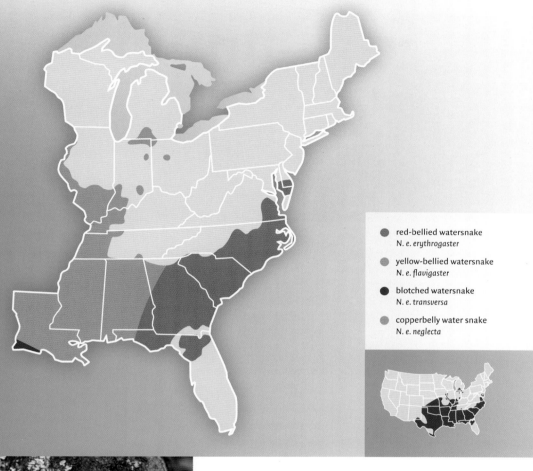

- red-bellied watersnake
  N. e. erythrogaster
- yellow-bellied watersnake
  N. e. flavigaster
- blotched watersnake
  N. e. transversa
- copperbelly water snake
  N. e. neglecta

Baby plain-bellied watersnakes have patterns on the back and sides that closely resemble those of northern or southern banded watersnakes, but they have an unmarked belly.

What Do the Babies Look Like? Babies have banding patterns that closely resemble those of northern watersnakes but can be identified by their unmarked belly.

Distribution and Habitat Plain-bellied watersnakes are found in 16 eastern states. The red-bellied watersnake (*N. e. erythrogaster*) is found from southeast-

A red-bellied watersnake from Georgia

ern Alabama and the western panhandle of Florida to southeastern Virginia, Delaware, and Maryland; the yellow-bellied watersnake (N. e. flavigaster) is found from central Georgia through Louisiana and up to western Tennessee, Kentucky, and Illinois; the blotched watersnake (N. e. transversa) enters into southwestern Louisiana from the west. The copperbelly watersnake (N. e. neglecta) occurs in scattered localities in southern Michigan, central Ohio, southern Indiana, Illinois, and Kentucky. Because of their propensity to travel long distances over land, plain-bellied watersnakes can turn up in or around any aquatic habitat or even in areas long distances from water. They are commonly associated with rivers and floodplains, large and small lakes, and ponds, including fish hatcheries and other aquatic habitats. In some regions of Indiana the copperbelly watersnake is most likely to be found in shallow, often temporary wetlands, ponds, and sluggish backwaters of river swamps.

Behavior and Activity Plain-bellied watersnakes usually hibernate during the coldest periods in winter but frequently become active earlier in the spring than other watersnakes in their area. They can be found throughout the warmer months basking on the banks or the vegetation along rivers, streams, and open bodies of water. They are more likely to be found on land than any other watersnake, presumably because they are traveling between wetland areas. They are active during the day in most seasons and at night during the summer months in the southern part of their range.

Food and Feeding Plain-bellied watersnakes eat primarily frogs of all sizes, which they capture both in the water and on land. They are also opportunistic predators on salamanders and fish, and will feed ravenously on larval amphibians when they are present in wetland habitats. They actively forage for prey and will trap tadpoles and fish against their body or vegetation in shallow water. They sometimes wave their mouth around and snag swimming prey on their teeth. Like other watersnakes, they capture and swallow prey alive without using constriction.

Reproduction Mating occurs from April through mid-June. Females give birth to live young, usually during August or September. Although exceptionally large litters have been reported (55 babies from one female in North Carolina), the typical litter size is about 18 (range, 2–55).

Predators and Defense Plain-bellied watersnakes have many predators, both terrestrial and aquatic, because of their tendency to move over land between aquatic habitats. Documented predators include largemouth bass, kingsnakes, cottonmouths, egrets, red-shouldered hawks, and red-tailed hawks. They typically react to threats as the other large watersnakes do—by trying first to escape into the water but, if captured, then biting and releasing musk if the threat persists. In contrast to other watersnakes, though, these snakes will sometimes leave the water and attempt to escape on land.

A red-bellied watersnake from North Carolina

Like the other large watersnakes, plain-bellied watersnakes have stout bodies and broad heads.

left The belly of yellow-bellied watersnakes is usually pale yellow. *above* In contrast to adults of the other eastern subspecies of plain-bellied watersnakes, adult blotched watersnakes have alternating blotches on their back and sides.

Conservation Federal legislation does not protect the plain-bellied watersnake in any of the southern states, but Indiana, Michigan, and Ohio have all listed the species as endangered. In 1997 the northern populations of the copperbelly watersnake subspecies in Michigan, Ohio, and northern Indiana were listed as threatened under the federal Endangered Species Act, and populations in Illinois and Kentucky were also given special protection. Their propensity for overland movement increases their chances of encountering highways and subsequent road mortality. Wetland degradation can also threaten this species because individuals often depend on more than one aquatic habitat during the active season.

What's in a Name? J. R. Forster named this snake *Coluber erythrogaster* in 1771 based on the Greek words *erythro*, meaning "red," and *gaster*, meaning "belly." Forster, who never visited North America himself, based the description on Captain J. B. Bossu's book about France's Louisiana Territory, which Forster translated into English. Bossu's outrageous account of the species, which he also called "the whipper," states that "it is sometimes about 20 feet long" and when it finds someone in the water "it twines round him so violently as to take away his breath, and drown him." In 1949 Roger Conant described the yellow-bellied watersnake (*N. e. flavigaster*), deriving its name from Latin *flavus*, meaning "yellow," and the copperbelly watersnake (*N. e. neglecta*), whose name derives from Latin *neglecta*, indicating the lack of attention the subspecies had received. The name of the blotched watersnake (*N. e. transversa*), which was described by Philadelphia physician Dr. Edward Hallowell in 1852, is from the Latin word *transversa*, meaning "from side to side," referring to the banding pattern on the body that remains in this subspecies into adulthood. Some herpetologists do not recognize any of the subspecies of *Nerodia erythrogaster* as distinctive.

BROWN WATERSNAKE *Nerodia taxispilota*

SCALES
Keeled

ANAL PLATE
Divided

BODY SHAPE
Heavy bodied; broad, thick, triangular head

BODY PATTERN AND COLOR
Dark brown, with square blotches on the back alternating with smaller blotches on the sides

DISTINCTIVE CHARACTERS
Eyes set more on top of the head than on the sides

SIZE

Description This heavy-bodied watersnake has a light brown back with dark brown, squarish blotches down the middle alternating with, but usually not connecting to, blotches on the sides. Large adult females often appear solid brownish gray or mud colored. The yellow or cream belly has irregularly spaced dark spots. The head is distinct from the body and is triangular. The eyes are set relatively higher on the head than those of most snakes.

What Do the Babies Look Like? The babies resemble the adults but may be more boldly marked.

top A brown body with darker squarish blotches on the back that alternate with smaller blotches on the sides is characteristic of the brown watersnake.

BROWN WATERSNAKE *Nerodia taxispilota*

Brown watersnakes frequently bask in trees and shrubs along waterways.

Distribution and Habitat Brown watersnakes are found from southeastern Virginia through most of the Coastal Plain of the Carolinas south through Florida and west to southern Alabama. They live in large rivers and adjacent swamps, streams, and reservoirs created by damming rivers or streams. They are not as prevalent in other wetland habitats within their geographic range and are seldom found far from the water.

Behavior and Activity Brown watersnakes hibernate in the northern parts of their range but may remain active during the winter in southern regions. Studies indicate that they may hibernate underwater. Along waterways where they are common they may be seen on the banks during the day in warm weather basking on rocks or on the limbs of trees and bushes. They usually position themselves directly above the water, into which they retreat when disturbed. They actively forage at night and during the day.

Food and Feeding Brown watersnakes feed almost exclusively on fish, especially catfish, which made up well over half of the reported food items documented in several studies of this species. Like diamondback watersnakes, their close relatives, brown watersnakes sometimes use an unusual ambush foraging tactic in which the snake coils its tail around a branch in the water and waits for fish to come within striking range. At other times they actively swim around,

Brown watersnakes vary little in color or pattern throughout their range. Like other watersnakes, female brown watersnakes get much larger than males.

Brown watersnake swimming in Everglades National Park, Florida

probing underwater holes and crevices in search of prey. Occasionally, a brown watersnake is found with the lateral (pectoral) spines of an ingested catfish protruding from its sides. The spines eventually decompose, and the snake apparently recovers with little difficulty.

Reproduction Brown watersnakes mate in the spring, and one to three males may accompany a single female and attempt to mate with her. Males apparently find females by following their pheromone trails. Mating snakes have been observed on the ground and in trees overhanging water. Large females give birth to up to 61 babies, but the typical litter size is about 20–30. Babies are usually born on land near the water in the late summer or fall.

Predators and Defense Because of their frequent association with large streams and rivers, brown watersnakes have a wide variety of natural predators, including hawks, large fish, cottonmouths, and alligators. They usually retreat quickly to the water when disturbed, but if captured will readily bite and release rather foul-smelling musk.

Conservation Like the other large watersnakes that occur within or near the geographic range of the cottonmouth, brown watersnakes are occasionally killed in the mistaken belief that they are venomous. They may be extremely abundant along some rivers in the Southeast, although pollution or development along rivers that reduces or degrades their natural habitats poses a threat to them.

What's in a Name? John Edwards Holbrook described this species in 1838 from two specimens, one from the "sea-board of South Carolina," the other from "the neighbourhood of the Altamaha River in Georgia." The species name is derived from the Greek words *taxis*, meaning "arrangement," and *spilos*, meaning "spots" in reference to the alternating blotches on the body.

DIAMONDBACK WATERSNAKE *Nerodia rhombifer*

SCALES
Keeled

ANAL PLATE
Divided

BODY SHAPE
Heavy bodied, broad head

BODY PATTERN AND COLOR
Brown to grayish green with dark, somewhat diamond-shaped, chainlike markings

DISTINCTIVE CHARACTERS
Eyes set more on top of the head than on the sides

SIZE
baby
typical
maximum
0'   3'   6'

Description Diamondback watersnakes are stout to very stout snakes with a series of dark, somewhat diamond-shaped markings that often connect on the back alternating with smaller markings on the sides. The background color is usually brown, gray, or olive-gray. The markings typically form a light, chainlike pattern running the length of the snake. The belly is usually cream to yellow with brown or black spots.

What Do the Babies Look Like? The babies look like the adults but are more boldly marked and the belly is often more orange with black spotting.

Distribution and Habitat Diamondback watersnakes in the eastern states range from central Alabama and the western tip of the Florida panhandle through

top The markings of a diamondback watersnake usually form a chainlike pattern running the length of the snake, as seen in this individual from Kentucky.

Mississippi and Louisiana to the western portion of Tennessee and Kentucky and the southwestern tip of Illinois. They can occur in large numbers in almost any habitat with standing or moving water, including small ponds, rivers, large lakes, and reservoirs. They are frequently seen around spillways of lakes, presumably in search of the small fish that abound in such areas.

> **DID YOU KNOW?**
>
> *Contrary to popular myth, snakes can bite underwater. Watersnakes and cottonmouths feed on fish, which they capture with their mouths underwater.*

Behavior and Activity Diamondback watersnakes are generally inactive in the winter but may become temporarily active during warm spells. During cold or extremely hot periods they retreat to hiding spots in or near water such as beneath root masses, under logs, in the burrows of other animals, and beneath overhanging banks of rivers or lakes. They are primarily active in the water at night and often can be found basking on limbs and other aquatic vegetation during the day.

Food and Feeding Diamondback watersnakes eat a wide variety of fish, including catfish, eels, sunfish, mullet, and mosquitofish, as well as frogs and toads. Adults eat mostly catfish. They sometimes feed by swiping the open mouth through the water and snapping at anything they happen to touch. They also exhibit the same behavior as brown watersnakes, wrapping the tail around a limb and dangling the body into the water to catch passing fish. These snakes have been observed in incredibly high densities at fish hatcheries and feeding on discarded fish carcasses near marinas.

Reproduction Like other watersnakes, diamondback watersnakes give birth to live young. Large females sometimes produce litters of more than 50 young, but the typical litter size is about 25. In the spring, males apparently find females by following pheromone trails, and mating may occur on land or in water, often with more than one male attempting to mate with a single female. The babies are born in the late summer or early fall.

Predators and Defense Diamondback watersnakes commonly fall prey to the natural predators typically found in and around lakes, rivers, and reservoirs, including raccoons, cottonmouths, large catfish, alligators, and wading birds. They are noted for biting savagely and discharging a nauseating musk when captured. Given the chance, however, they will always try to escape a predator (including humans) first.

DIAMONDBACK WATERSNAKE *Nerodia rhombifer*

Catfish are a common prey item of diamondback watersnakes.

Diamondback watersnakes frequently bask on logs and other debris.

Conservation Diamondback watersnakes are common in many aquatic habitats. Although harmless to humans, their superficial similarity to cottonmouths and their co-occurrence with that species in many parts of their geographic range result in large numbers being killed by people in some areas.

What's in a Name? This species was described as *Tropidonotus rhombifer* in 1852 by Dr. Edward Hallowell based on specimens "found abundantly on the borders of streams" in the vicinity of the Arkansas River. The species name is derived from the Latin word *rhombifer*, meaning "rhomb-bearing," which refers to the diamond-shaped pattern on the back. Only one of the three subspecies occurs in the United States, the other two being restricted to Mexico.

EASTERN GREEN WATERSNAKE *Nerodia floridana*

SCALES
Keeled

ANAL PLATE
Divided

BODY SHAPE
Heavy bodied

BODY PATTERN AND COLOR
Gray or greenish with yellowish or cream-colored belly

DISTINCTIVE CHARACTERS
Row of small scales between eye and upper lip scales

SIZE

Description This heavy-bodied watersnake is solid grayish or greenish above with a yellow, unmarked belly that becomes darker under the tail. The body color in southern Florida is often reddish. A row of scales (suboculars) encircles the lower half of the eye, separating it from the upper lip scales; among eastern snakes, only the western green watersnake has a similar arrangement.

What Do the Babies Look Like? The babies are similar to adults but with faint crossbands.

Distribution and Habitat Eastern green watersnakes are found throughout most of Florida, sporadically in southern Georgia, and in parts of the southern half of South Carolina in open, marshy wetlands with minimal tree cover such

top Eastern green watersnakes are grayish or greenish above.

EASTERN GREEN WATERSNAKE *Nerodia floridana*

Eastern green watersnakes from southern Florida often have a reddish body color.

as Carolina bays. They are not typically residents of rivers, streams, or swamps, but do sometimes occupy reservoirs.

Behavior and Activity Like other large watersnakes of the eastern United States, eastern green watersnakes hibernate in the colder northern parts of their range but are active year-round in southern Florida. Even in areas with cold temperatures they often bask on sunny days in late winter, and if conditions remain warm they may stay active throughout the spring, summer, and fall. They frequently travel over land, especially in southern Florida during summer rains.

Food and Feeding Little is known about the diet of eastern green watersnakes. Most food records are of fish, including sunfish, bass, and crappies; pig frogs; and salamanders, including sirens. Nothing is known of their hunting techniques, but they presumably employ an active foraging strategy. Like other watersnakes, they swallow their prey alive.

Reproduction The details of reproduction are poorly known. A few matings have been observed in the late winter or early spring. Females give birth in the

This eastern green watersnake from the Savannah River Site in South Carolina is from the northernmost known population of the species.

Eastern green watersnakes in Florida are frequently found crossing roads that intersect wetlands adjacent to the highway, especially during rains or at night.

summer, sometimes to huge litters. The record litter is 132 babies, which were dissected from a dead female, but the typical litter size is much smaller, about 20–40 young.

Predators and Defense As is the case with other animal species that have large numbers of offspring, most baby eastern green watersnakes die before reaching adulthood. Predators characteristic of their wetland habitats include river otters, wading birds, hawks, ospreys, turtles, kingsnakes, alligators, and predatory fish. If unable to escape, eastern green watersnakes usually bite and release musk from scent glands in the cloaca.

Conservation Eastern green watersnakes are very abundant in some aquatic ecosystems; however, their distribution in the northern parts of their range is patchy, and more information is necessary to determine their true status there. Occasionally, people kill eastern green watersnakes intentionally, and thousands die annually on Florida highways adjacent to wetlands. They are also sensitive to wetland habitat disruption resulting from urbanization and development.

What's in a Name? C. C. Goff initially described the eastern green watersnake in 1936 as a subspecies of the green watersnake (*Nerodia cyclopion floridana*), the name denoting the state of Florida, where most of the specimens used for the original description were collected. Based on the absence of markings on most of the belly, Goff considered the snakes to be distinctive from western examples of what was then recognized as the same species. He described the body color of some of the specimens from southern Florida as "copper-brown." D. D. Pearson elevated the eastern green watersnake to a full species, distinct from the western green watersnake, in 1966. Some herpetologists call this species Florida green watersnake, although it is also found in Georgia and South Carolina.

WESTERN GREEN WATERSNAKE *Nerodia cyclopion*

SCALES
Keeled

ANAL PLATE
Divided

BODY SHAPE
Heavy bodied

BODY PATTERN AND COLOR
Gray or greenish with mottled gray belly

DISTINCTIVE CHARACTERS
Row of small scales between eye and upper lip scales

SIZE

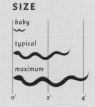

Description Western green watersnakes are generally solid grayish, greenish, or olive above, sometimes with faint dark bands, and have a dark belly with irregularly spaced yellowish spots. Like the eastern green watersnake, this species has a row of scales encircling the lower half of the eye (suboculars) and separating it from the upper lip scales.

What Do the Babies Look Like? Babies are similar to adults but with more distinctive crossbands.

Distribution and Habitat Western green watersnakes occur along the Gulf Coast from the western tip of the Florida panhandle into Texas and up the Mississippi River drainage through the western portions of Tennessee and Kentucky to extreme southwestern Illinois. They are common residents of swamps,

top Western green watersnake from southern Illinois

Among North American watersnakes, only the western (shown here) and eastern green watersnakes have scales that encircle the eye and separate it from the upper lip scales.

bayous, ditches, open marshes, and small lakes and ponds. They have been reported from areas with brackish water along the Gulf Coast.

Behavior and Activity Western green watersnakes become inactive during the winter wherever and whenever cold temperatures occur. They generally hibernate under vegetation, in holes along the bank, or in stumps or rotten logs close to wetland habitats. They seldom venture far from water. They are active in the daytime during the spring and fall and active primarily at night during the summer months. Individuals frequently can be seen basking on logs and limbs above water. Western green watersnakes commonly travel over land between wetlands and are often found crossing roads.

Food and Feeding These apparently opportunistic aquatic foragers feed on many species of fish and amphibians. Little is known of their foraging strategy, but presumably they actively search out prey using scent as their main method of detection. Prey items are seized and swallowed alive.

Reproduction Little is known about this snake's reproduction. Most of the evidence points to mating in the spring, and many males may try to court a single receptive female. Females presumably announce their readiness to mate by releasing pheromones detectable by males. Young are born in the summer, and litters range in size from 6 to 37 (average, about 17).

Predators and Defense Western green watersnakes can fall victim to any of the typical predators inhabiting southern swamps, lakes, and other wetlands, including raccoons and river otters, wading birds, cottonmouths, kingsnakes, and alligators. They bite viciously when captured by another animal or a person, and like other watersnakes release a foul-smelling musk.

Conservation Besides habitat destruction and degradation, the greatest threats to western green watersnakes are vehicles on highways passing through swampy areas and intentional killing because of negative attitudes about snakes.

WESTERN GREEN WATERSNAKE *Nerodia cyclopion*

What's in a Name? A. M. C. Duméril, G. Bibron, and A. H. A. Duméril described this species in 1854 in the book *Erpétologie générale ou Histoire Naturelle complète des Reptiles*. The species name is derived from the Greek words *kycklos*, meaning "circle," and *ops*, meaning "eye," referring to the scales encircling the eye that include the small suboculars between the upper lip scales and the eye itself. Among North American watersnakes this scale configuration is present only in eastern and western green watersnakes, which were once considered a single species. Some herpetologists refer to this species as the Mississippi green watersnake, although it is found in nine other states.

MUD SNAKE *Farancia abacura*

SCALES
Smooth

ANAL PLATE
Usually divided

BODY SHAPE
Stout, with rounded head

BODY PATTERN AND COLOR
Shiny black above; belly with black-and-red checkerboard pattern

DISTINCTIVE CHARACTERS
Sharp spine on end of tail

SIZE
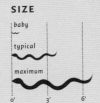

Description Mud snakes are heavy-bodied, shiny black, and typically have a red-and-black checkerboard pattern on their belly. The red belly coloration extends onto the sides of the body in some individuals. In some cases, particularly on younger snakes, the red coloration may extend far up onto the sides of the body, appearing as thin bands, especially anteriorly. The rounded head is not very distinct from the neck, and there is a harmless spine on the tail tip. Rarely, individuals have white pigment instead of red. Two subspecies, the eastern mud snake (*F. a. abacura*) and the western mud snake (*F. a. reinwardtii*), have been described and intergrade over a broad geographic area.

top Mud snakes are shiny black above, usually have partial red bands, and typically have a red-and-black checkerboard pattern below.

A widespread anomaly seen in mud snakes is that the red coloration is replaced with white.

Compared with bands of adult mud snakes the red bands of babies often extend much higher up from the belly.

What Do the Babies Look Like?

Baby mud snakes look like the adults, but the red markings sometimes extend farther up onto the body, sometimes forming complete red bands on the neck.

Distribution and Habitat

Mud snakes range throughout the southeastern Coastal Plain from southern Virginia through all of Mississippi and Louisiana, up the Mississippi River drainage through western Tennessee and Kentucky to southwestern Illinois. Adults occupy a variety of aquatic habitats, including cypress swamps, ponds, oxbow lakes, river floodplains, and even isolated wetlands—wherever their primary prey, amphiumas (large aquatic salamanders), are present. Young mud snakes are found in aquatic habitats near the area where they hatched, including smaller wetlands where amphiumas may be absent but tadpoles and salamander larvae are abundant.

Behavior and Activity Mud snakes take refuge from cold weather by crawling into burrows made by other animals as well as under leaf litter, rotten logs, dead palmetto fronds, and other ground debris in terrestrial habitats near aquatic areas. They may be active at any time of the year, including warm periods in the winter. Mud snakes are active on land and in the water both at night and during the day throughout much of the warm part of the year. Recently hatched young are commonly found on land near wetlands in the early spring.

Food and Feeding Mud snakes specialize on amphiumas ("congo eels") but will also eat sirens, which are also large, aquatic salamanders; juveniles prey on

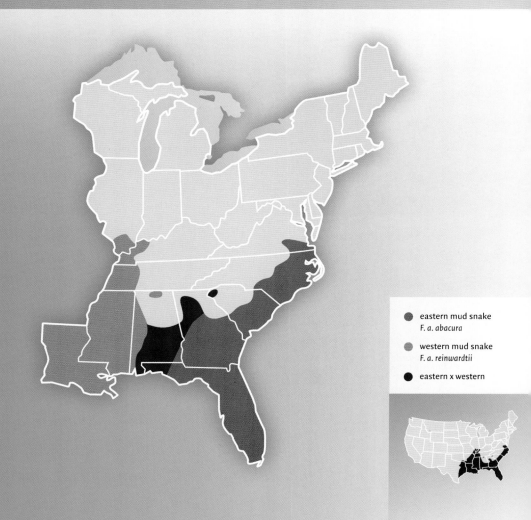

MUD SNAKE *Farancia abacura*

- eastern mud snake *F. a. abacura*
- western mud snake *F. a. reinwardtii*
- eastern x western

various species of aquatic salamanders, frogs, tadpoles, and occasionally fish. The enlarged teeth in the rear of the mouth and sharp spine on the end of the tail may help the snake hold onto slippery prey.

> **DID YOU KNOW?**
>
> *The mud snake is one of more than 20 eastern snakes that typically do not bite people, even when handled.*

Reproduction Mud snakes apparently mate in the spring or early summer. The females lay their eggs in rotting vegetation, under logs, or in similar places during late spring or summer. Clutches contain 4–111 eggs, with an average of about 30. Females often curl around their eggs and stay with them until they hatch in late summer or early fall. Baby mud snakes do not enter the water upon hatching but instead spend their first winter on land.

Predators and Defense Known predators include cottonmouths, alligators, and barred owls. Large fish as well as terrestrial carnivores such as raccoons, racers, and hawks also presumably prey on mud snakes. When threatened, some mud snakes display the red undersurface of the tail in a spiral, possibly to divert attention away from the head. Captured mud snakes will often coil around the captor's arm and harmlessly poke the tail spine against the person's skin. Mud snakes do not bite people, but they may release a foul-smelling musk.

The primary prey of adult mud snakes are giant salamanders such as the greater siren being consumed here.

Mud snakes, such as this one from Illinois, seldom venture far from water.

Conservation Among the greatest threats to mud snakes are the destruction and degradation of wetland habitats and the areas where these interface with the upland nesting habitats.

What's in a Name? The mud snake was the last species described in the 1836 publication *North American Herpetology; or the Reptiles Inhabiting the United States*, volume 1, by John Edwards Holbrook of Charleston. He described the snake as "rare and shy" and noted that he had seen them only in South Carolina. The genus name *Farancia*, given by John Edward Gray of the British Museum in 1842, was apparently a nonsense word that Gray made up. One explanation for *abacura* as the species name is that *abacus* means "square stone" in Latin and is descriptive of the distinctive pattern of black and red squares on the belly. The western mud snake subspecies *F. a. reinwardtii* was described in 1837 by Hermann Schlegel from a specimen from Louisiana and named in honor of Caspar G. C. Reinwardt, a Dutch botanist noted for explorations of the East Indies. Mud snakes are known as hoop snakes in some parts of the South because of an old superstition that they roll down a hill in the shape of a hoop and then stab a person with the tail spine.

RAINBOW SNAKE *Farancia erytrogramma*

SCALES
Smooth

ANAL PLATE
Usually divided

BODY SHAPE
Robust body, rounded head

BODY PATTERN AND COLOR
Black above with three thin red stripes; chin yellow; belly red with black dots

DISTINCTIVE CHARACTERS
Spine on end of tail; red stripes on back

SIZE
baby
typical
maximum
0'   3'   6'

Description Rainbow snakes are shiny, iridescent black, heavy-bodied snakes with three thin red stripes running the length of the back. The belly is mostly red or pinkish with two or more rows of large black dots. The red, and yellow when it is present, often extend up onto the sides. The head is round and indistinct, and the tail has a harmless spine. The chin and neck are often yellow. Two subspecies, rainbow snake (F. e. erytrogramma) and south Florida rainbow snake (F. e. seminola), have been described.

What Do the Babies Look Like? The young look like miniature adults.

top A shiny black body with three red stripes and a yellow chin make a rainbow snake easily identifiable.

The underside of a rainbow snake is distinctive in coloration and pattern.

The tail spine of a rainbow snake is harmless to humans and is presumably used in helping to secure its slippery prey of American eels.

Distribution and Habitat Rainbow snakes are found throughout most of the Atlantic and Gulf Coastal Plain from southern Maryland to the southern half of Mississippi and into Louisiana. Adults are found in association with streams, river systems, swamps, and other aquatic habitats inhabited by their primary prey, American eels. They characteristically nest in upland areas with sandy soil, and the young spend their first year or two in smaller isolated wetlands that have an abundant supply of tadpoles and salamander larvae.

Behavior and Activity Rainbow snakes are inactive during the winter, but both young and adults become active by mid-spring. Hibernation and warm-weather refuge sites are presumably beneath living root masses, in rotten stumps and logs, and in animal burrows or decayed root holes in and around river swamps. They forage primarily at night in the water but will also capture eels during the day and have been known to move long distances on land away from water. Rainbow snakes that are ready to shed are occasionally seen out of the water basking on floating logs along stream or river banks.

Food and Feeding The principal prey of adult rainbow snakes is American eels; juveniles eat a variety of aquatic salamanders and tadpoles. As is the case with

RAINBOW SNAKE Farancia erytrogramma

- rainbow snake
  F. e. erytrogramma
- south Florida rainbow snake
  F. e. seminola

The primary prey of adult rainbow snakes is the American eel.

their close relatives the mud snakes, the tail spine may help the snake hold onto the slimy prey.

Reproduction Little is known of the mating habits or other aspects of the reproductive biology of rainbow snakes. They are believed to mate in the spring; the eggs are laid in underground nests in sandy soil during the summer. Clutches range in size from 10 to 52 eggs. The eggs hatch during late summer or fall, but the young do not enter the aquatic environment until the following spring.

> **DID YOU KNOW?**
> After the south Florida rainbow snake was declared extinct by the U.S. Fish and Wildlife Service, the Center for Snake Conservation continued to conduct intensive searches to determine if they still exist. Thus far, none have been found.

Predators and Defense Like mud snakes, rainbow snakes probably fall victim to both terrestrial and aquatic predators. They are quite docile when captured by humans and do not bite.

Conservation Habitat preservation can be critical to the long-term persistence of this species. Stream or river alterations and dams or commercial overfishing that affect the abundance of American eels, their primary food, will have direct impacts on rainbow snakes. Human modifications, including highways that create barriers or increase mortality of adult females traveling from lowland swamps to upland habitats to lay eggs, could severely reduce rainbow snake populations in an area. The south Florida rainbow snake, known only from Glades County, has not been found since the mid-1960s and has been officially declared extinct by the U.S. Fish and Wildlife Service.

What's in a Name? This snake was described and named *Coluber erytrogrammus* by the French naturalist Palisot de Beauvois in 1801 based on a specimen from "North America" that was later determined to be from "the lower Cooper River, in the vicinity of Charleston, SC." The species name is derived from the Greek *erythro*, meaning "red," and *gramma*, meaning "written," referring to the red lines on the body. In 1964 Wilfred T. Neill described a rainbow snake captured in Fisheating Creek in Glades County, Florida, as a subspecies (*F. e. seminola*; south Florida rainbow snake) for which he suggested the common name Seminole rainbow snake in honor of the Seminole tribe of Florida. Neill captured an additional specimen in the same habitat and observed another preserved one from the same county "near Lake Okeechobee."

venomous snakes

previous page A copperhead from North Carolina

Venomous snakes are among the most maligned and misunderstood animals on earth. Perhaps only sharks and wolves have generated more misconceptions and misinformation. To really understand venomous snakes, a person must view them in a practical and objective manner. Venomous snakes have fangs and venom for capturing prey, not for biting people. Because no venomous snake preys on humans, the only reason they bite humans is in self-defense, and nearly always only as a last resort when they feel cornered or threatened.

Unfortunately, venomous snakes have a reputation for being aggressive toward humans, but in fact, aggression has nothing to do with how venomous snakes respond to human encounters. Cottonmouths are an excellent example. They have a reputation among the general public, and even among some scientists, for being extremely aggressive. Many people believe cottonmouths will actually pursue them with the intent of biting. In fact, a study done in South Carolina provided strong evidence that cottonmouths are very reluctant to defend themselves by biting. In that study, conducted in the swamps along the Savannah River, investigators tested how more than 50 cottonmouths responded to being stepped on and picked up by humans (note: the investigators used snake-proof boots to do the "stepping on" and an artificial arm to do the "picking up"). When first encountered, the cottonmouths usually tried to remain unnoticed or tried to escape. When stepped on, they usually tried to bluff by gaping their mouth widely, vibrating their tail, and releasing musk, but only rarely (less than 5 percent of the time) did they bite the investigator's boot. Even when picked up, only about 35 percent of the snakes

Pit vipers, such as this young cottonmouth, have elliptical pupils.

Venomous snakes are generally very reluctant to bite people. Most cottonmouths will warn an intruder with their mouth open, but fangs not extended, rather than bite.

bit the artificial arm, and most of those snakes were already aggravated from having been stepped on! Many other animals that people never associate as "aggressive" toward humans would be much more likely to bite under the same circumstances. Imagine, for example, the results of first stepping on and then picking up a squirrel! The results of this study should help to dispel a common misconception about a species of venomous snake unjustifiably perceived as aggressive.

Only 7 of the 67 species of snakes found in the eastern United States are venomous: the copperhead, cottonmouth, coral snake, and four species of rattlesnakes. No single species occurs over the entire eastern United States, and most areas have fewer than four venomous snake species. Additionally, the rarity of some species in many areas makes it extremely unlikely that a person will ever encounter them.

VENOMOUS OR POISONOUS?

Many people refer to venomous snakes as poisonous snakes. In fact, the word "venomous" is more appropriate than "poisonous" because of the way the toxin is administered to the victim. Venom is a toxic substance produced in special glands by one animal that is injected into the body of another animal mechanically, such as by stinging or biting. A poison, in contrast, is a substance that is toxic when consumed or touched. Thus, pit vipers and the coral snake are better referred to as venomous because they inject toxins into other animals through their hollow fangs.

BIOLOGY OF VENOMOUS SNAKES

A rattlesnake rattle is composed of loosely connected segments. A new one is added at the base of the string each time the snake sheds its skin.

The cottonmouth, copperhead, and rattlesnakes are all pit vipers (family: Viperidae, subfamily: Crotalinae). Vipers have movable fangs in the front of the mouth that fold up against the roof of the mouth when not in use. Pit vipers are named for their "pits," special heat-sensing organs that look like a hole between the eye and nostril. The pits sense infrared radiation, such as heat produced by warm-blooded prey. All vipers in the Americas are pit vipers, including tropical species such as the bushmaster and fer-de-lance. The venom of most pit vipers contains hemotoxic proteins that destroy blood and tissue, although the venom's toxicity varies among snake

The tiny rattle of a pigmy rattlesnake often sounds like an insect.

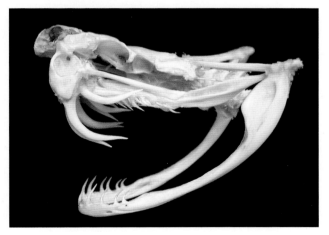

The skull of a rattlesnake showing the movable fangs and replacement fangs

species. Diamondback rattlesnakes, for example, have fairly potent hemotoxic venom, but the venom of copperheads is generally less so.

The coral snake is the only venomous snake native to the eastern United States that is not a pit viper. The coral snake is a member of the family Elapidae, which includes the cobras, kraits, and mambas. Members of this family have relatively short fangs in the front of the mouth that do not fold back like those of vipers, and they lack heat-sensitive pits to detect warm-blooded prey. Their venom is primarily neurotoxic and acts by inhibiting the nervous system—including the nerves that control breathing. Prey animals that have been bitten usually die because they are unable to breathe.

SNAKEBITE

One of the biggest misconceptions about snakes is the extent of the danger they pose to humans. A person in the United States is far more likely to be killed by lightning than to die from a venomous snake's bite. Certainly, some snakes can seriously harm or even kill humans. But although hundreds of people are bitten each year by copperheads (which are responsible for most of the venomous snakebites in the eastern United States), very few human deaths have been documented. Detailed medical records are difficult to obtain, but the American Association of Poison Control Centers reported no deaths from cottonmouths, copperheads, or eastern coral snakes over a 20-year period from 1983 to 2003.

The venom of coral snakes is highly toxic and attacks the nervous system.

 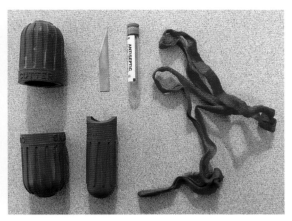

left The best snakebite kit is a set of car keys, a cell phone, and a companion to get the victim to an emergency treatment facility. *right* You should NEVER use the old snakebite kits that require cutting and use of a tourniquet.

According to the Centers for Disease Control and Prevention, approximately 7,000–8,000 people are bitten by venomous snakes in this country each year (of which an average of about 5 are fatal) while domestic dogs bite more than 4.5 million people (average annual fatalities number more than 15). Like many other wild animals, as well as domestic pets such as dogs and cats, venomous snakes can be dangerous. But snakes are also natural and important parts of our native ecosystems and natural heritage.

Using common sense is the best way to minimize your chance of getting bitten by a venomous snake. Become familiar with the snakes in your area. Never try to capture or handle a venomous snake—or any snake of unknown identity. Children should be taught to appreciate snakes but to leave all of them alone unless they are with an adult who can reliably identify the species occurring in their region and distinguish between the venomous species and harmless ones. When you are outside in areas where venomous snakes occur, use a flashlight at night, wear sturdy shoes, and watch where you step and put your hands.

A bite from a venomous snake should be treated as a medical emergency. First, try to remain calm. Remember, the probability of dying from a venomous snakebite in the United States is extremely low. With proper medical treatment, most people recover fully and with few complications. Seek medical attention as quickly as possible and call your state's poison control center. Doctors at poison control centers often have considerably more experience treating snakebite than do doctors at hospitals. DO NOT try to cut the snake-bitten area. DO NOT use a tourniquet. DO NOT drink alcohol. DO NOT apply ice to the area

In some areas, venomous snakes are recognized as potentially dangerous but are still appreciated as important parts of our natural heritage.

or use electric shock treatment. All of these are likely to do more harm than good. Most snakebite kits do little, if anything, to lessen the severity of snakebite, and some, such as the old kits that promoted cutting and use of a tourniquet, are likely to cause harm. If a child is bitten by a venomous snake, the consequences could be more serious simply because the same quantity of venom will have a proportionally greater impact on someone with a smaller body size. The good news is that a study of more than 1,000 venomous snakebites in the United States found that in more than 60 percent of cases, the snake injected an inconsequential amount of venom or sometimes no venom at all. Consequently, what may appear to be a life-threatening situation may actually turn out to be medically insignificant. Nevertheless, the safest approach is to get anyone, child or adult, bitten by a venomous species to an emergency care facility as quickly as possible, in addition to contacting your state's poison control center.

COPPERHEAD *Agkistrodon contortrix*

SCALES
Keeled

ANAL PLATE
Single

BODY SHAPE
Heavy bodied, but less so than other pit vipers

BODY PATTERN AND COLOR
Light brown body with darker, hourglass-shaped crossbands

DISTINCTIVE CHARACTERS
Top of head coppery brown

SIZE

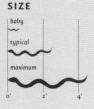

Description Copperheads have dark brown, hourglass-shaped crossbands on a light brown background. The crossbands may be broken in places or offset on either side of the body. The top of the head is usually coppery brown but in some individuals may be reddish or orange. The belly is pale brown with irregularly spaced darker markings. The background color of the southern copperhead (A. c. *contortrix*) is rather light and sometimes looks pinkish. The northern copperhead (A. c. *mokasen*) has wider crossbands than the southern subspecies and is usually darker overall. The

The copperhead gets its name from its coppery head color.

top A light brown body with dark, hourglass-shaped crossbands is characteristic of the copperhead.

COPPERHEAD *Agkistrodon contortrix*

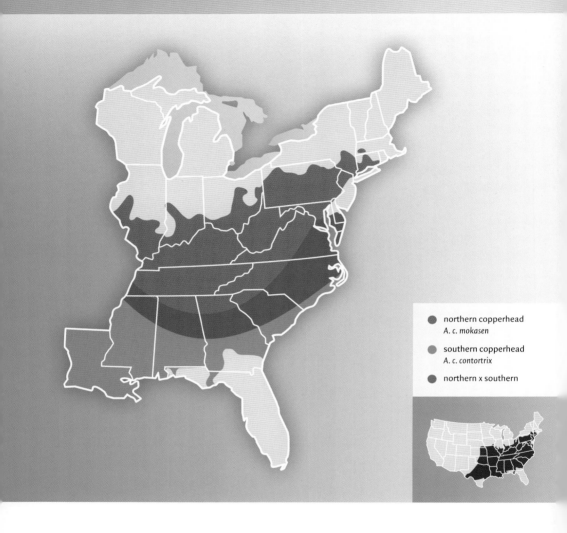

- northern copperhead
  A. c. mokasen
- southern copperhead
  A. c. contortrix
- northern x southern

two subspecies intergrade over a very broad geographic area. Copperheads have a broad, triangular shaped head characteristic of other pit vipers. However, many nonvenomous species, especially some watersnakes, expand their head when threatened so that it looks very similar to that of a pit viper. Hence, head shape alone is not a good indicator of whether a snake is harmless or venomous.

What Do the Babies Look Like? The babies have the same pattern as the adults but have a yellow or green tail tip.

A northern copperhead from Indiana

Distribution and Habitat Copperheads are wide ranging and may be very common in areas where they occur. They are absent in the South from the southeastern one-third of Georgia and most of Florida except for a small portion of the panhandle in the Apalachicola River drainage. In the North their range extends into the southern portions of Illinois, Indiana, Ohio, Pennsylvania, the southeastern tip of New York, northern New Jersey, and southern and central New England. They are found in all types of terrestrial habitats, including rocky areas in the mountains; mixed hardwood and pine forests; swamp margins; and even farms, suburban areas, and coastal island resorts if sufficient prey is available and ground vegetation or other cover is present for concealment.

Behavior and Activity Copperheads are noted for hibernating in communal dens with several to dozens of other individuals, but such behavior is most characteristic of northern and mountain populations. They some-

Baby copperheads use their bright yellow tail as a lure to attract potential prey such as frogs or lizards.

Southern copperheads from the sandhills regions of South Carolina often have a lighter body color.

times hibernate communally with timber rattlesnakes as well as nonvenomous species. Southern Atlantic and Gulf Coastal Plain copperheads are more likely to hibernate alone, but occasionally several can be found together in a suitable hibernation site. Stumps with old root holes, rocky crevices, and other underground sanctuaries are common retreats. Copperheads spend many days of the year aboveground, either coiled in a setting that provides camouflage, such as leaf litter, or moving about. They are generally more active at night during the summer months but may be found in the daytime during cool periods in spring and fall.

Food and Feeding Copperheads are opportunistic predators that eat a wider variety of prey than most other venomous snakes. The diet of adults consists mostly of vertebrates such as voles, mice, birds, frogs, lizards, and other snakes, but they will eat large insects too. They usually wait in ambush for their prey but sometimes actively forage and have been observed climbing trees to capture cicadas. Juveniles enhance their ability to ambush prey by wiggling their yellow tail tip, which attracts small lizards and frogs within striking range. A copperhead feeding on a large prey animal such as a rodent strikes the prey and immediately releases it, then tracks the dying animal by following its odor trail. Other prey, such as frogs and birds, are struck and held until the venom takes effect.

Reproduction Copperheads have two mating seasons: late spring and early fall. After mating, the females can store sperm in special receptacles for long periods before fertilization occurs, which is typically in the spring. A male copperhead finds a mate by following her pheromone trail with his highly sensitive forked tongue and then actively courts her by touching her with his snout and rubbing her neck with his chin. When a male copperhead encounters another male during the mating season, they may engage in a "combat dance" in which each tries to pin the other to the ground. The young are born in the late summer or early fall. Typical litter size is 7 or 8, but very large females may produce more than 20 babies. In the mountains, female copperheads congregate and give birth prior to entering communal hibernation dens. Most females give birth every 2 years, taking the alternate year to build up enough energy reserves to have more offspring, but they may reproduce annually if enough food is available.

Predators and Defense Small copperheads presumably fall prey to any animal that can avoid or will risk a mildly venomous bite. Adults are somewhat less vulnerable, but common kingsnakes, which are immune to the venom, will eat copperheads as readily as other prey. Copperheads respond to kingsnakes by lifting the center of the body upward to ward off attack rather than by attempting to strike. Copperheads' first line of defense is always camouflage or concealment, and they are masters at it. People come within striking distance of innumerable copperheads each year but pass by unaware. Once discovered, a copperhead will usually vibrate its tail and strike if it feels threatened. A strike toward a person may be from several feet away with no chance of successfully biting and is assumed to be a form of threat display. Copperheads release a musky odor when disturbed, and most will bite if they are picked up.

Conservation Copperheads are not considered to be threatened by environmental degradation in most areas because they can live in a variety of habitats, including wooded neighborhoods and other areas with moderate urbanization. Populations in northern parts of their range are declining, however, and they are protected in several states. Nonetheless, individual copperheads are often persecuted by humans who think the snake poses a danger to themselves or others. Unfortunately,

> **DID YOU KNOW?**
>
> *A common misconception is that the venom of juvenile snakes is more potent than that of adults. This is not necessarily true, and furthermore, young venomous snakes cannot inject nearly as much venom as adults.*

Recently born copperheads usually remain together until they shed for the first time.

many harmless snakes, such as ratsnakes, corn snakes, milksnakes, and northern and southern banded watersnakes, are killed each year because of their superficial resemblance to copperheads.

How Dangerous Are They? More people are bitten by copperheads in the eastern United States than by any other venomous snake. The bite can be very painful, but very few deaths have ever been documented. The venom destroys red blood cells and other tissues but in most instances is not considered highly toxic to humans. However, if you are bitten, you should call your state's poison control center and seek medical attention immediately. If you find a copperhead in the field, you can observe it from a safe distance but should not approach it.

What's in a Name? Carl Linnaeus described this species in 1766 and named it *Boa contortrix*, stating "Habitat in Carolina." Palisot de Beauvois in 1799 provided the genus name *Agkistrodon*, derived from the Greek word *ankistron*, meaning "fishhook," and *odontos*, meaning "tooth," referring to the curved fangs. Although *contortus* means "twisted" and *rix* is a suffix meaning "one who performs," the application to a description of this species is equivocal. The name for the northern subspecies, *A. c. mokasen*, is derived from the Algonquin word for the copperhead. This species is referred to as the highland moccasin in some areas.

COTTONMOUTH *Agkistrodon piscivorus*

SCALES
Keeled

ANAL PLATE
Single

BODY SHAPE
Heavy bodied, some large individuals extremely so

BODY PATTERN AND COLOR
Black to olive-brown, sometimes with dark bands

DISTINCTIVE CHARACTERS
Black line from eye to back of lower jaw

SIZE
baby
typical
maximum
0'   2'   4'

Description Cottonmouths have wide, dark crossbands on a brown to olive-brown background, but some large individuals may be almost solid black. The belly has irregularly spaced, dirty yellow and brown blotches, and the underside of the tail is usually black. The eastern cottonmouth (A. p. piscivorus) generally has a lighter background color and has no pattern on its snout. The western cottonmouth (A. p. leucostoma) looks like the eastern cottonmouth but is usually darker. The Florida cottonmouth (A. p. conanti) is usually very dark to solid black and has two dark vertical bars on its nose. Cottonmouths that live in water with high concentrations of tannins can develop a solid coppery color. A broad zone of intergradation occurs among the subspecies.

top Adult cottonmouths usually have wide, dark crossbands on a brownish background, but some may be almost solid black.

The coloration and markings of a baby cottonmouth (left) and a copperhead can be very similar, but the cottonmouth's crossbanding is less well-defined.

What Do the Babies Look Like? Baby cottonmouths are lighter in color, often reddish brown, and more boldly patterned than adults and have a yellow to greenish tail tip. Baby cottonmouths are frequently misidentified as copperheads.

Distribution and Habitat Cottonmouths occupy at least part of every coastal state from southern Virginia to Louisiana but are absent from most of the Piedmont and mountainous areas. Their range extends northward through western Tennessee and Kentucky, southwestern Illinois, and extreme southern Indiana. They are associated with river swamps, backwaters, and floodplains throughout their geographic range, but are less common on the rivers themselves. They are prevalent around lakes, reservoirs, small ponds, streams, and isolated wetlands, especially when such habitats are surrounded by swampy habitat or heavy vegetation.

Behavior and Activity Cottonmouths hibernate during cold winter weather, but in southern states a few individuals can often be found aboveground on sunny winter days, usually coiled near a hole into which they can readily escape. Hibernation retreats may be on land several hundred feet from water and are often in stump holes or holes left by decayed roots, under logs, or in abandoned animal burrows. Beaver lodges seem to be particularly attractive to cottonmouths in fall and winter, probably because of the many hiding places

COTTONMOUTH *Agkistrodon piscivorus*

- eastern cottonmouth
  A. p. piscivorus
- Florida cottonmouth
  A. p. conanti
- eastern x Florida
- western cottonmouth
  A. p. leucostoma
- eastern x western

A cottonmouth has a distinctive black line from the eye to the back of the lower jaw.

Cottonmouths typically swim with their head elevated and their body above the surface of the water.

they offer both in and out of the water. In most of Florida, cottonmouths are active year-round. During the active season, cottonmouths move around both in the daytime and at night, commonly hunting after dark during hot summer weather. They can be found at any time of day coiled on stumps, logs, or land associated with river swamps and other wetland habitats. On cool spring and fall days they climb into the lower branches of trees or bushes to bask. When traveling, cottonmouths are as likely to be seen on land as in the water, but they are seldom found far from aquatic habitats.

Food and Feeding Cottonmouths are opportunistic predators that eat a wide assortment of prey, which they capture using both ambush and active foraging methods. They are even known to scavenge fish and roadkills. Their diet includes small mammals, birds, fish, frogs, salamanders, turtles, other snakes (especially watersnakes), and even baby alligators. They also occasionally eat invertebrates. The cottonmouth usually strikes and holds its prey, though particularly dangerous animals such as rodents that might bite or scratch may be released and then tracked by their scent. Juvenile cottonmouths attract prey the way young copperheads do, by wiggling their yellow tail tip while waiting in ambush.

Reproduction Cottonmouths mate primarily in the spring but also in late summer and fall. Males find females by following their pheromone trails. Like many other pit vipers, males "fight" with competing males in a "combat dance" to win the privilege of mating with available females. Female cottonmouths can store sperm for long periods. Litters of 1–20 babies (average, about 7) are born in late summer or early fall. Several female cottonmouths will sometimes congregate when giving birth. A female cottonmouth typically has young at intervals of every 2–3 years.

Predators and Defense Among the most successful natural predators on cottonmouths are alligators, kingsnakes, and larger cottonmouths. Large mammals, wading birds, hawks, crows, and owls are also known or suspected predators, but most of these species restrict themselves to juveniles because of the potentially greater hazard associated with attacking adults. Cottonmouths rely on camouflage and concealment for defense and will usually remain motionless unless they are certain they have been discovered. The first response of many individuals to a perceived threat is to try to escape, either down a nearby hole, under a stream bank, or into the water if time allows. If a safe and successful escape seems uncertain, most cottonmouths will stand their ground. The common name of the species is derived from the open-mouthed threat display that is characteristic of many, but not all, individuals. The lining of the mouth is white rather than the darker pink, red, or even black typical of many other species of snakes. Other threat displays or defensive behaviors include rapidly vibrating the tail, flattening the head and body to look larger, and releasing a strong-smelling musk. Cottonmouths sometimes release musk before they are actually approached and can be located by the smell.

A cottonmouth from southern Illinois displays the trademark threat display of the species.

Behavioral research with cottonmouths in the field and laboratory has demonstrated conclusively that they are not aggressive toward humans; their behavioral responses are solely defensive in nature. These studies have also shown that most cottonmouths are extremely reluctant to bite, and many refuse to even if picked up. Nevertheless, if you are walking in an area where cottonmouths occur, be careful where you step.

Cottonmouths often congregate on beaver dams or logs in river swamps where they can safely retreat if threatened.

Conservation Cottonmouths are not threatened as a species, but localized populations may be extirpated when wetland habitats are destroyed or highways are built between the aquatic habitats where cottonmouths feed and the upland sites where they hibernate.

How Dangerous Are They? If you find a cottonmouth in the field, leave it alone! Cottonmouths can be dangerous if harassed, but they can be safely observed from a distance of several feet. Take care when walking in known cottonmouth habitat such as swamps and wetlands. A snake may not move if it thinks it has not been seen, posing a potential hazard for someone who unwittingly steps on one. The venom destroys blood cells and has anticoagulant properties, and bites have resulted in fatalities, but nearly all victims who receive proper medical treatment survive. If bitten, you should call your state's poison control center and get medical attention immediately.

What's in a Name? B. G. E. Lacepède is credited with the original description of this species in 1789. The species name, *piscivorus*, means "fish eater." The western subspecies (*A. p. leucostoma*) was described by Gerard Troost in 1836. *Leuco* means "white" in Latin and *stoma* means "mouth," referring to the snake's white-mouthed defensive display. Lacepède's original description did not mention the white mouth coloration, probably because he worked only with preserved specimens and thus was not aware of the display, whereas Troost actually collected the cottonmouths he described. The Florida cottonmouth was named as a subspecies (*A. p. conanti*) by Howard Kay Gloyd in 1969 in honor of Roger Conant, noted herpetologist and director of the Philadelphia Zoo. Some herpetologists consider the Florida cottonmouth to be a full species, although it is similar in appearance to cottonmouths in other parts of the range. Despite attempts by herpetologists, state game officials, and conservationists to declare that the common name for this species is cottonmouth, the names water moccasin and cottonmouth moccasin are deeply ingrained throughout much of the South.

PIGMY RATTLESNAKE *Sistrurus miliarius*

SCALES
Keeled

ANAL PLATE
Single

BODY SHAPE
Heavy bodied

BODY PATTERN AND COLOR
Gray, grayish brown, or red with dark blotches

DISTINCTIVE CHARACTERS
Very tiny rattle; enlarged plates on head

SIZE
baby
typical
maximum
0'   2'   4'

Description Pigmy rattlesnakes are usually gray or grayish brown with a line of dark blotches—and frequently also a gold or orange stripe—running along the middle of the back. They usually have at least one row of smaller dark blotches running along each side, a dark stripe from the eye to the jaw, and two irregular dark stripes on top of the head. The light-colored belly has irregular dark blotches. Pigmy rattlesnakes have a tiny rattle that is often very difficult to hear. The Carolina pigmy rattlesnake (S. m. miliarius) is gray and has a well-defined head pattern. Individuals from Hyde and Beaufort Counties in North Carolina have a reddish background color. The head pattern is obscured in the dusky pigmy rattlesnake (S. m. barbouri), while the head pattern of the western pigmy rattlesnake (S. m. streckeri) is well defined. The rattle segments are fragile and

top Most pigmy rattlesnakes are gray or grayish brown with dark blotches.

PIGMY RATTLESNAKE *Sistrurus miliarius*

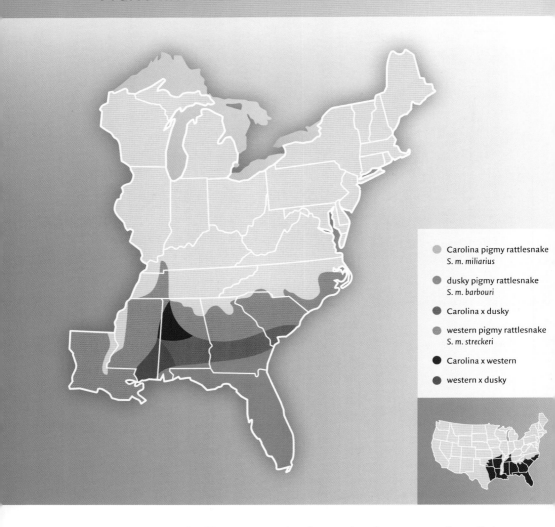

sometimes break off, so some snakes have only a few or no rattles. All three subspecies intergrade with one another in zones of contact.

What Do the Babies Look Like? Baby pigmy rattlers look like the adults but have a yellow-tipped tail.

Distribution and Habitat Pigmy rattlesnakes are found throughout Florida and in parts of other southeastern states from North Carolina west to Tennessee and southwestern Kentucky, and south to Louisiana. They vary regionally in their habitat choices, ranging from dry upland sandhills and mixed hardwood and pine forests in some areas to low-lying, sometimes flooded palmetto stands, floodplains, mangroves, and marshy habitats in others.

Behavior and Activity Pigmy rattlesnakes hibernate or at least become inactive during the winter in most parts of their geographic range, but even during periods of cold weather some individuals may be found basking on sunny days. Seasonal peaks of activity vary. In many parts of the Coastal Plain, autumn is the period of highest aboveground activity, but pigmy rattlesnakes are also encountered frequently in early-to-late spring in many areas. Movement over land is more common at night but may also occur during the day. In areas where their

top left Pigmy rattlesnakes from North Carolina around Pamlico Sound are often reddish.   *top right* A Carolina pigmy rattlesnake from Georgia   *bottom* Dusky pigmy rattlesnakes are often dark colored.

habitat becomes flooded, pigmy rattlesnakes will often climb several feet above the ground into bushes, vines, or palmettos.

Food and Feeding Pigmy rattlesnakes feed on a wide variety of small animals including lizards, snakes, frogs, centipedes, and mice. Adults capture prey primarily by ambush, and juveniles may use their yellow-tipped tail to lure prey within striking range. Pigmy rattlesnakes use chemical cues when selecting ambush sites. They generally strike their prey, release it, and then track the dying animal.

Reproduction Pigmy rattlesnakes are known to mate in the fall and store sperm until the following year. Females usually reproduce every other year. Typically 6 or 7—but occasionally more than 30—babies are born in August or September. Females from the northern parts of the range (e.g., North Carolina) generally give birth to fewer young than those in the south (e.g., Florida). Newborn pigmy rattlesnakes are tiny and generally stay their first week of life with their mother, who may defend them.

Predators and Defense Despite being venomous, pigmy rattlesnakes become prey themselves for many other species because of their small size. Any large snake-eating snake (e.g., kingsnakes, racers, and indigo snakes) can eat them, as can medium to large mammals, hawks, and owls. Large frogs, toads, shrews, and larger ground-feeding birds such as turkeys may eat the babies, although the bite of even a small pigmy rattlesnake would deter many potential predators. When coiled in pine straw, palmettos, or other natural habitats, these little snakes are exceedingly difficult to see, even by those skilled at finding them. Thus, their response to an encounter with a human is first to remain motionless if in a coiled position. A pigmy rattlesnake threatened while crawling over the ground will generally coil and face its adversary. In most

> **DID YOU KNOW?**
>
> Pigmy rattlesnakes have proportionally the smallest rattle of any species of rattlesnake.

The rattles on even an adult pigmy rattlesnake are tiny.

316 Venomous Snakes

Like most pit vipers, baby pigmy rattlesnakes usually stay with their mothers for a week or more after birth.

instances they will not strike unless pestered, and usually bite only if picked up—and sometimes not even then. No venomous snake should ever be picked up, however, because all species have the potential to bite and inject venom.

Conservation Because they are so well camouflaged, pigmy rattlesnakes are probably more common than is apparent in places where they occur. Many populations are probably unknowingly lost to habitat development or to timber management and agricultural techniques that chop or till the soil and remove vegetation. Many pigmy rattlesnakes are killed crossing roads each year.

How Dangerous Are They? Because of their small size, pigmy rattlesnakes are generally not considered particularly dangerous to humans. Most bites occur when someone picks up a snake or steps on it, and bites in Florida are common. Nearly all victims who receive proper medical treatment recover fully. If you find a pigmy rattlesnake in the field, observe it from a safe distance but do not attempt to capture it or disturb it in any way.

What's in a Name? Carl Linnaeus called this species *Crotalus miliarius* in the 12th edition of *Systema Naturae*, published in 1766, identifying the specimen as being from "Carolina." The origin and significance of *miliarius* is uncertain. In 1883 Samuel Walton Garman placed the pigmy rattlesnake and massasauga in a separate genus, *Sistrurus*, a name derived from a Latin word meaning "rattle." The dusky pigmy rattlesnake (*S. m. barbouri*) and the western pigmy rattlesnake (*S. m. streckeri*) are named after two well-known U.S. herpetologists: Thomas Barbour and John Kern Strecker Jr. Pigmy rattlesnakes are known as ground rattlers in many areas.

MASSASAUGA *Sistrurus catenatus*

SCALES
Keeled

ANAL PLATE
Single

BODY SHAPE
Heavy bodied

BODY PATTERN AND COLOR
Gray or grayish brown with dark blotches

DISTINCTIVE CHARACTERS
Enlarged plates on head

SIZE

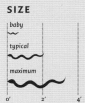

Description Massasaugas are usually gray or grayish brown with large dark blotches on the back and smaller ones on the sides that touch the scales on the belly. The back and side blotches connect with each other on the tail to form dark rings. Occasional individuals have a very dark body in which the blotching pattern is not accentuated. A broad dark stripe bordered by two whitish stripes runs back from the eye to the back of the jaw. The belly is solid black or black with white or yellow markings. Massasaugas have a smaller rattle than most rattlesnakes, but not nearly as small as that of their close relatives the pigmy rattlesnakes.

top Most massasaugas have gray or grayish-brown bodies with large dark blotches on the back and smaller ones on the sides. The blotches connect with each other on the tail to form dark rings.

MASSASAUGA *Sistrurus catenatus*

What Do the Babies Look Like? Baby massasaugas look like the adults.

Distribution and Habitat The eastern massasauga (S. c. *catenatus*) is found from western New York and Pennsylvania across the central and northern portions of Ohio, Indiana, and Illinois; in southern Wisconsin; and throughout the Lower Peninsula of Michigan. According to Gary Casper of the University of Wisconsin—Milwaukee, the historical range of the species in Wisconsin was once much larger and more contiguous than its cur-

DID YOU KNOW?

The chemical composition of venom of juvenile pit vipers may differ from that of adults if young snakes feed on different prey than their parents.

rent distribution. Massasaugas are generally associated with marshy habitats, often adjacent to wooded areas or open fields.

Behavior and Activity Because of their northern distribution, massasaugas hibernate during the winter throughout their range, often in deep crayfish burrows or stump holes. They become active by April and do not become dormant again until the onset of cold weather in October. They are noted for moving relatively long distances in some situations between marsh habitats and upland areas while foraging during the active season, but the most extensive movements are by males searching for females in late summer. Individuals studied in Wisconsin, Illinois, New York, and Pennsylvania ranged over 5 or more acres in a season.

Food and Feeding Massasaugas primarily feed on small rodents such as voles and jumping mice, but they are opportunistic and have been documented to prey on a wide diversity of vertebrate species, including small mammals, birds and their eggs, snakes and lizards, frogs, and fish. They also eat crayfish, insects, and other invertebrates.

A baby massasauga from Grayling, Michigan, coiled in typical habitat. Note that the tail of juvenile massasaugas is not yellow like that of the pigmy rattlesnake.

Reproduction Eastern massasaugas mate predominantly in late summer. Litter size averages around 8 (range, 2–20) with the young typically being born a year later in August and September. Male-male combat in which two males "wrestle" each other trying to pin the other to the ground has been observed during the mating season.

Predators and Defense Known massasauga predators include amphibians (bullfrogs), snakes (racers), birds (loggerhead shrikes), and mammals (weasels and coyotes). Raptors (hawks and owls) and other carnivorous mammals in the region are also likely predators. Like other pit vipers, their first line of defense is camouflage, but if challenged they will rattle, face their attacker, and strike and bite to defend themselves.

Conservation Massasaugas have been eliminated from many localities by urbanization, commercial development, and agriculture that degrade or destroy the wetlands they require. Because they often concentrate at hibernation sites, they are vulnerable to collectors who remove them or others intent on killing

Some adult massasaugas develop very dark bodies in which the blotched pattern is not obvious.

them. The species is recognized to be in peril throughout its range and is a candidate for federal listing as threatened.

How Dangerous Are They? Because of their small size, massasaugas are considered less dangerous to humans than the two species of large eastern rattlesnakes. Although they have less venom for a defensive bite, however, it is nonetheless potent. Human deaths resulting from massasauga bites are extremely rare. As with other U.S. venomous snakes, most bites occur when someone picks up the snake or steps on it, and nearly all victims who receive proper medical treatment recover fully. Finding a massasauga in the field is a notable wildlife experience, but the snake should only be observed from a safe distance with no attempt to capture or disturb it.

What's in a Name? Constantine Samuel Rafinesque described this species in 1818, giving the scientific name *Crotalinus catenatus* to what he called the chained rattlesnake. He selected the Latin word *catena*, which means "chain mail armor," for the species to indicate its appearance. The common name massasauga is believed to originate from a Chippewa word referring to the mouth or delta of a large river, a habitat characteristic of the species.

TIMBER/CANEBRAKE RATTLESNAKE Crotalus horridus

SCALES
Keeled

ANAL PLATE
Single

BODY SHAPE
Heavy bodied

BODY PATTERN AND COLOR
Gray, brown to almost black, or yellow with dark chevrons

DISTINCTIVE CHARACTERS
Black tail with rattles at end

SIZE
baby
typical
maximum
0'   4'   8'

Description Timber and canebrake rattlesnakes typically have black chevrons or crossbands on a gray, brown, or yellow background. The belly is yellow to grayish white and stippled with black dots. Throughout most of their geographic range, especially in mountainous areas, these big rattlesnakes are called timber rattlesnakes. Those found in the Coastal Plain are usually called canebrake rattlesnakes. Although the common names differ according to the locality, timber and canebrake rattlesnakes are a single species. Certain characteristics are generally typical of one form or the other, however, and it is often useful to refer to them separately. Timber rattlesnakes usually have a yellowish brown or gray background but are sometimes completely black. Canebrake rattlesnakes are lighter in color, and the background coloration often has a pinkish hue. Cane-

top A solid black tail is characteristic of timber and canebrake rattlesnakes.

The yellow phase of a timber rattlesnake from Macon County, North Carolina

Some timber rattlesnakes are entirely black.

brake rattlesnakes also have an orange or brown stripe running the length of the back. Both forms have a solid black tail ending in a distinctive set of large rattles.

What Do the Babies Look Like? The babies are replicas of adults except they are more gray when first born and have a single rattle (button) at the end of the tail.

Distribution and Habitat Historically the species was found in parts of all but two eastern states, Michigan and Delaware, and was absent from most of the Florida panhandle and peninsula, and from portions of North Carolina, Virginia, and Louisiana (see map). Although this continues to be a geographically widespread species, populations have been extirpated in some regions over large areas where they once occurred. Populations apparently no longer exist in Maine and Rhode Island. Most of the populations in the other New England states are also gone. Rattlesnakes have disappeared from New York City and Long Island, and their geographic ranges have contracted appreciably in many areas of Upstate New York. Likewise, most timber rattlesnake populations in Ohio, Indiana, and Illinois are now extinct, with most remaining localities being in the southern one-third of these states. In North

TIMBER / CANEBRAKE RATTLESNAKE *Crotalus horridus*

Baby timber rattlesnakes stay with their mother for several days after being born.

Carolina and Virginia they have been extirpated from much of the Piedmont as a result of human activity.

Timber and canebrake rattlesnakes occupy virtually all terrestrial habitats within their geographic range, including hardwood forests, pine flatwoods, rocky mountainous terrain, floodplains, high ground in swamps, and rural agricultural areas. Their persistence, however, becomes less likely with human occupation, and few remain in urbanized or residential areas, or even in areas with substantial agriculture.

Behavior and Activity These rattlesnakes spend a few weeks in the Southeast to a few months in northern and mountainous areas hibernating each winter, depending on the climate of the region and the severity of the winter weather. Those in mountainous areas are noted for hibernating in communal dens with copperheads and other snakes. Canebrake rattlesnakes in most of the Coastal Plain hibernate singly, although an ideal den site such as a large stump hole with deep root tunnels or a rocky area with many crevices may attract several individuals. During active months they find temporary refuges in animal burrows, culverts, or under pieces of tin and wood around abandoned homesites. These rattlesnakes generally become active in mid-to-late spring and may remain active into late fall. Both timber rattlesnakes and canebrake rattlesnakes can be found moving around during the day, although they commonly move at night during hot weather. Timber rattlesnakes may travel up to 6 miles from den sites to foraging habitats or in search of mates, and often return in the fall to the same overwintering sites. Less is known about the annual movement patterns of canebrake rattlesnakes, but presumably they can also move long distances. Large males searching for females during the late summer and fall mating season move the greatest distances. A male canebrake rattlesnake tracked by radiotelemetry over a 1-month period in the Pine Barrens of New Jersey covered a straight-line distance of almost 7 miles from its summer feeding and mating habitat to a hibernation site.

> **DID YOU KNOW?**
>
> Female timber rattlesnakes may take 10 years to reach maturity and may reproduce only every 3–4 years.

Food and Feeding Timber and canebrake rattlesnakes eat small mammals such as shrews, mice, voles, rats, squirrels, and rabbits almost exclusively, and may rarely consume birds. Like most other rattlesnakes, they are ambush hunters that sit motionless for extended periods waiting for prey to approach. Ambush points are usually at a location with strong prey odors, such as along a log or

Timber rattlesnakes in Pennsylvania congregating around a sunny rocky outcrop

other places frequented by small mammals. Some individuals hunt by sitting at the base of a tree with their head pointing up, waiting for a squirrel to come down the trunk. They usually strike and bite the prey, inject venom, and then release it and follow the dying animal's scent until it can be safely eaten.

Reproduction Female timber and canebrake rattlesnakes may take 5–10 years to mature and are capable of reproducing every 2–3 years thereafter, although some may wait as long as 6 years in times of low food availability (e.g., during a drought). Timber and canebrake rattlesnakes mate primarily in the late summer and fall, though some breeding may occur in the spring. After mating, females store sperm in their oviducts over the winter and use it to fertilize their eggs in the spring. Litters of 6–18 are produced in late summer (canebrake rattlers) or early fall (timber rattlers). Female timber rattlesnakes in the mountains often congregate around open, sunny rocky outcrops. These "rookeries" provide good opportunities for thermoregulation during the final stages of gestation and serve as protected sites for giving birth. Females usually stay with their babies for up to 2 weeks, until the first time they shed their skin. In the mountains, recently born young will follow the scent trail of their mother or other timber rattlesnakes back to the winter den.

Predators and Defense Adults have few natural enemies because most potential predators tend to avoid them. Kingsnakes are immune to the venom of rattlesnakes, however, and probably attempt to eat those they encounter. Kingsnakes and indigo snakes can kill and eat rattlesnakes that are almost the same length as they are. Carnivorous mammals probably prey on young rattlesnakes, and hawks are known to attack even large timber rattlesnakes. Rattlesnakes of all sizes may be susceptible to mammalian predators during the winter when their lower body temperatures leave them incapable of defending themselves effectively. Timber and canebrake rattlesnakes are usually relatively benign when humans encounter them. Their first response is usually to become motionless, remaining coiled, or stretched out if they were moving. Many will not even rattle when approached, and people have actually stepped on large rattlesnakes without being bitten. If pestered, they will usually rattle, coil, and assume a defensive pose, and most will bite if approached too closely or picked up.

Conservation Timber and canebrake rattlesnakes throughout their range are threatened by habitat destruction and development, especially by the construction of highways that they must cross to reach seasonal habitats. The widespread disappearance across much of the species' former geographic range in the eastern United States is especially due to habitat degradation and human encroachment into natural habitats. Roads are also particularly hazardous for this species. Of 200 canebrake rattlesnakes examined in a South Carolina study, 84 percent were found on highways, usually dead, especially during the fall breeding season.

> **DID YOU KNOW?**
>
> *You cannot age a rattlesnake by the number of its rattles. Rattlesnakes add a rattle segment every time they shed their skin, which may be several times each year. Also, rattle segments break off the end of the rattle string from time to time.*

How Dangerous Are They? Timber and canebrake rattlesnakes are potentially very dangerous and should be treated with the utmost respect. Although they are generally very reluctant to bite people, when bites do occur they can be very serious, and some have resulted in death. The venom destroys blood and tissues;

A canebrake rattlesnake typically has an orange or brown stripe down the back.

A bite from a timber rattlesnake can be very dangerous, but most are reluctant to bite humans and will retreat if given the opportunity.

some populations, such those in the Pine Barrens of New Jersey and the coastal areas of South Carolina and Georgia, have been reported to have venom with neurotoxic elements that affect the nervous system, making bites from them even more life threatening. Continued research will be necessary to confirm the level of severity of bites from these localities and to determine the chemical composition of the venom. If you come across a rattlesnake in the field, you can watch it from a safe distance, but never approach too closely or attempt to harass or capture the snake.

What's in a Name? The timber rattlesnake has the distinction of being the first species of snake from North America to receive a formal scientific name. Carl Linnaeus named the species in 1758 on page 214 in the 10th edition of his classic book *Systema Naturae*, which introduced the consistent use of two Latin words to identify species of plants and animals. Under the entry for *Crotalus horridus* Linnaeus stated, "Habitat in America." The word *krotalon* means "rattle" in Greek, and the name *horridus* means "frightful" in Latin. Both were appropriate choices for naming this snake with its large fangs, menacing appearance, and string of rattles on the tail. Timber and canebrake rattlesnakes have been considered to be separate subspecies (timber rattler, *C. h. horridus*; canebrake rattler, *C. h. atricaudatus*) by numerous herpetologists. The Latin words *atri*, meaning "black," and *caudatus*, meaning "tail," refer to the black tail, and the canebrake rattlesnake is sometimes called the velvet-tailed rattler.

EASTERN DIAMONDBACK RATTLESNAKE *Crotalus adamanteus*

SCALES
Keeled

ANAL PLATE
Single

BODY SHAPE
Heavy bodied

BODY PATTERN AND COLOR
Brown, gray, or yellow background with black, light-bordered diamonds

DISTINCTIVE CHARACTERS
Two light stripes running from eye to jaw

SIZE

Description Eastern diamondback rattlesnakes are very heavy-bodied snakes with large, dark, diamond-shaped markings outlined with white on a brown, gray, or yellowish background. The diamond markings often fade toward the tail. The unmarked belly is pale gray to white. Eastern diamondback rattlesnakes have large rattles and two light stripes on either side of the head that start in front of and behind the eye and extend to the corner of the jaw.

What Do the Babies Look Like? Baby eastern diamondbacks are miniature versions of the adults.

Distribution and Habitat Eastern diamondbacks have the most limited geographic range of the eastern pit vipers, historically inhabiting much of the lower Coastal Plain from North Carolina to Louisiana in pine flatwoods and

top Eastern diamondback rattlesnake from Camden County, Georgia

> **DID YOU KNOW?**
>
> All snakes can swim, even rattlesnakes. However, except for the aquatic species of snakes most stay on the surface of the water and do not dive.

savannas, longleaf pine sandhill habitats, low-lying palmetto stands, and coastal islands. They are often found in habitats with abundant gopher tortoise burrows where they may seek shelter.

Behavior and Activity Diamondback rattlesnakes become inactive during cold winter weather but do not ordinarily hibernate for long periods because prolonged cold periods are uncommon in their range. Stump holes with extensive networks of decayed root tunnels and gopher tortoise burrows are preferred retreats from extremely hot or cold temperatures. During the warmer months, diamondbacks spend much of their time coiled beneath palmettos, grass clumps, and other vegetation, where their camouflage renders them almost invisible. Overland travel occurs commonly during daylight hours, but nocturnal movement has also been observed. Individual diamondbacks may move a mile or more on some occasions.

Eastern diamondback rattlesnakes thrive in longleaf pine habitat that frequently burns.

Food and Feeding Adults eat mammals such as cotton rats, squirrels, and rabbits. Young snakes feed primarily on smaller mammals such as mice. Diamondbacks are ambush predators that may remain for weeks in one spot waiting for prey to come by. After striking and injecting venom, they release the animal and allow it to crawl away to die, then use their excellent odor detection abilities to locate it. Large adult diamondbacks, which feed primarily on large prey such as cottontail rabbits, may consume only four or five meals each year.

Reproduction Eastern diamondback rattlesnakes mate mostly in the late summer or early fall, although spring mating has been reported. Mated females

EASTERN DIAMONDBACK RATTLESNAKE *Crotalus adamanteus*

An eastern diamondback rattlesnake will sometimes hide its head beneath its body when threatened.

The classic defense pose of an eastern diamondback rattlesnake is to keep its neck in an S-shaped curve ready to strike while the tail rattles a warning.

can store viable sperm for long periods. Young are born during the early fall and can take several years to reach maturity. The litter size is typically about 12 but can be more than twice that. Females usually give birth inside a tortoise burrow or old stump hole, and the mother stays with the babies, apparently to protect them, for a few weeks until they shed their skin for the first time. Eastern diamondback females reproduce only every 2–3 years. Adult snakes may live more than 20 years.

Predators and Defense The largest diamondback rattlesnakes have few natural enemies, although big kingsnakes and indigo snakes may eat some adults. Young diamondbacks are eaten by racers and kingsnakes, and probably by large mammals such as raccoons, skunks, and foxes as well, although the cost of being bitten by even a small diamondback may be a deterrent to some predators. Diamondback rattlesnakes coiled in a camouflaged position often will not reveal their presence, but those physically disturbed or confronted by an adversary while crawling on the surface are unquestionably among the most dangerous snakes in America. A diamondback on the defensive typically lifts its head several inches above the ground, keeping the neck in an S-shaped curve ready to strike while the tail rattles a warning. Once aggravated, the snake continues to rattle, sometimes backing away into vegetation to escape. Although their actions are strictly a form of defensive behavior and not of aggression, diamondbacks can be provoked unwittingly, and one may bite before someone is aware of its presence, as in the rare instances when someone steps directly on one in the field.

Conservation Eastern diamondback rattlesnake populations appear to be doing well in Georgia and Florida where suitable habitat remains. However, diamondbacks do not usually persist in suburban or other developed areas and have apparently been eliminated from Louisiana and from much of their range in other states as well. Diamondbacks have minimal or no protection in most states, in part because of negative public attitudes and politicians' reluctance to provide protection for a venomous snake. Nonetheless, some conservation groups have recommended that the eastern diamondback rattlesnake be included on the list of federally endangered species.

Eastern diamondback rattlesnakes can be abundant in protected coastal areas.

Diamondback rattlesnakes are commonly collected in Georgia for commercial purposes. Whether rattlesnake roundups that still exist in Georgia and Alabama endanger rattlesnake populations continues to be debated. Promoters of such roundups claim that they do not affect rattlesnake numbers in the region, but reports based on long-term studies have noted potential impacts on populations, most notably a decrease in body size of adults over time. Some of the more notorious rattlesnake roundup events now call themselves "wildlife festivals" and offer educational programs. Some people still kill diamondback rattlesnakes even in wild areas where they pose little danger to humans, and public education appears to be the only way to ensure long-term preservation of this magnificent animal.

How Dangerous Are They? Eastern diamondback rattlesnakes are potentially among the most dangerous snakes in the United States, although bites are extremely rare. Their venom is highly toxic, and adults can inject much more of it than smaller venomous species can. The venom rapidly destroys tissues and blood cells, and bites not treated with antivenom have resulted in death within 24 hours. Fortunately, eastern diamondbacks are usually very reluctant to defend themselves against humans by biting except as a last resort or when they have been completely surprised. If you see one in the field, you should consider yourself fortunate (many herpetologists have never seen one) and watch it from a safe distance. Never harass or try to capture a diamondback rattlesnake.

What's in a Name? Palisot de Beauvois described this species from "the southern parts of the United States" in 1799. He referred to it as a water rattlesnake because "it lives near the waters." He stated that it was "non-descript" and that he had seen many that had been killed "for the sake of diminishing their number, and extracting the grease, of which an oil is prepared" to treat rheumatism. The Latin word *adamanteus* means "resembling a diamond" and refers to the "parallelograms or lozenges of a browner colour than the rest of the body" described by Palisot de Beauvois.

CORAL SNAKE *Micrurus fulvius*

SCALES
Smooth

ANAL PLATE
Usually divided

BODY SHAPE
Slender

BODY PATTERN AND COLOR
Red, yellow, and black rings

DISTINCTIVE CHARACTERS
Black nose; red and yellow rings adjacent

SIZE
baby
typical
maximum
0' 2' 4'

Description Coral snakes are slender, smooth-scaled snakes with red, yellow, and black rings. The red rings always in contact with yellow rings and the black snout distinguish coral snakes from similarly colored snakes such as the scarlet snake and scarlet kingsnake. The Texas coral snake (M. f. tener) has large amounts of black pigment on the red bands that form spots or blotches, whereas the eastern coral snake (M. f. fulvius) has only tiny spots of black on the red bands.

What Do the Babies Look Like? The babies look like miniature adults.

Distribution and Habitat Coral snakes can potentially be found anywhere in Florida but are patchily distributed in other southeastern states, where they are confined mostly to parts of the lower Coastal Plain from the Carolinas to

top Distinctive features of a coral snake are a black nose and red and yellow rings that are adjacent to each other.

CORAL SNAKE *Micrurus fulvius*

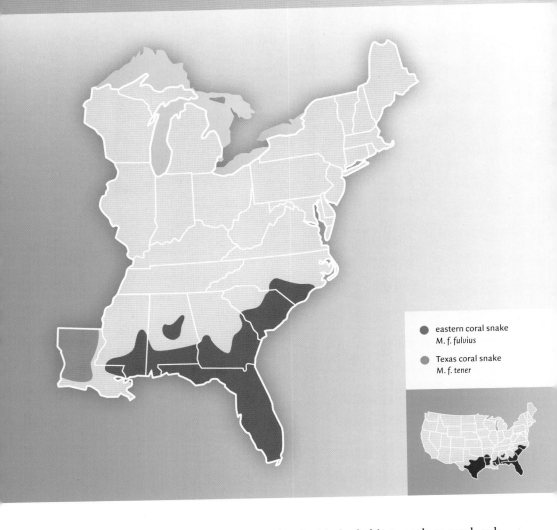

- eastern coral snake *M. f. fulvius*
- Texas coral snake *M. f. tener*

Louisiana. They are typically associated with dry habitats such as scrub oak–pine sandhills and, in Florida, hardwood hammocks, but they also thrive in frequently flooded pine flatwoods.

Behavior and Activity Coral snakes may hibernate for several weeks in the northern or inland portions of their geographic range but are inactive for shorter periods during winter in southern Florida and along the coast. They are characteristically most active in the spring and fall, especially during wet years, and are seen less commonly in the summer. They presumably spend most of their

Coral snake from Telfair County near Macon, Georgia

time underground in tunnels left by decayed roots, in burrows of small mammals and insects, and in other natural passageways beneath the surface. Overland travel is almost always during the daytime, and more often during the morning than at other times of the day. Coral snakes seldom if ever climb into trees or bushes.

Food and Feeding Coral snakes feed primarily on lizards and other snakes. They are active foragers that search for prey by poking their head under leaves and other debris and by following underground tunnels created by roots and burrowing animals. Prey are grabbed and chewed until venom is injected and begins to take effect. The venom kills by disrupting the functioning of the central nervous system. Coral snakes apparently use both visual and chemical cues to detect potential prey.

A Florida Keys coral snake with reduced yellow rings

Reproduction Coral snakes mate in the spring and probably also in the early fall. The courting male slowly rubs his head and body over the female until she indicates her readiness to mate by raising her tail. Coral snake eggs are proportionally more elongate than those of most other snakes and are laid under-

Coral snakes, when aboveground, are often active during the daytime.

ground, in leaf litter, or in other moist protected places during the late spring or summer. Clutches usually contain about seven eggs but may total more than a dozen, and the babies hatch in August or September.

Predators and Defense Coral snake predators include larger snake-eating snakes such as kingsnakes, indigo snakes, and racers, and even larger coral snakes. Birds of prey sometimes take coral snakes, but presumably most recognize the distinctive color pattern and avoid them. Coral snakes have been known to kill full-grown hawks imprudent enough to try to capture them. A coral snake's first response to danger is to try to escape underground or beneath ground litter, where it can quickly disappear from view. If unable to escape, some individuals will flatten the body, put the head beneath it, and then raise the tail as if it were the head. If a predator goes after the tail, the coral snake might be

> **DID YOU KNOW?**
>
> More than 20 of the snake species described before the twentieth century, including the venomous coral snake, were originally placed in the genus Coluber.

Coral snakes, like this one from Orlando, may be common in Florida but are secretive and not often seen.

able to escape or bite the unsuspecting attacker. Coral snakes readily bite if picked up or even approached too closely, but like other venomous snakes will escape to safety if they can. Coral snakes often chew on their victims, including people who pick them up, and can sometimes be yanked off before much or any venom is administered.

Conservation Like pigmy rattlesnakes, coral snakes are often present but not seen, and the species is common in many parts of Florida. However, many populations in Florida and other southeastern regions have probably been lost because of urbanization, agriculture, or other land management practices that destroy natural upland habitats. In North Carolina, the northernmost part of their range, they are nearly extirpated.

How Dangerous Are They? Coral snakes have highly toxic venom that attacks the nervous system, including the nerves controlling the breathing muscles

RED TOUCH YELLOW, KILL A FELLOW— RED TOUCH BLACK, FRIEND OF JACK

Some nonvenomous snakes have patterns similar to the coral snake and are considered coral snake mimics. Potential predators might perceive these snakes to be dangerous and thus avoid them.

In the eastern United States you can tell a nonvenomous species such as the scarlet kingsnake or the scarlet snake from the venomous coral snake by using the little rhyme above. The venomous coral snake has red bands that touch its yellow bands; the harmless species do not.

Important note: this rule does not work in Latin America, where there are many other species of coral snake and many other red, yellow, and black-ringed species.

Scarlet kingsnakes can be distinguished from coral snakes by black rings that touch red rings and by a red nose.

(e.g., diaphragm and rib muscles). Because of their small size, however, the amount of venom they inject is usually rather small, and a person receiving medical treatment in a timely fashion is unlikely to die. Bites are very rare and generally occur only when the snake is picked up. If you find a coral snake in the field, it can be safely watched from a short distance away as long as you do not try to pick it up or harass it. If you find a coral snake in your yard or some other area frequented by humans, you should have it relocated to a suitable nearby habitat by someone experienced in handling venomous snakes.

What's in a Name? This species, named in 1766, was the first North American venomous snake described by Carl Linnaeus that was not a pit viper, although he named it *Coluber fulvius*, not recognizing it as belonging to a different family. He gave the original specimen's location as "Carolina." The Latin word *fulvius* means "reddish yellow," a reference to the body color. S. F. Baird and C. Girard described the Texas coral snake (*M. f. tener*) in 1853 as a full species, noting that it was "much more slender" than the eastern coral snake, hence the species name from Latin *tenerum*, meaning "delicate." The Greek *mikros* means "small" and "oura" means "tail," thus the genus name *Micrurus* refers to the proportionally short tail. Some herpetologists consider coral snakes from western Louisiana a separate species (*M. tener*), as originally described. The coral snake is sometimes called the harlequin snake or candy cane snake.

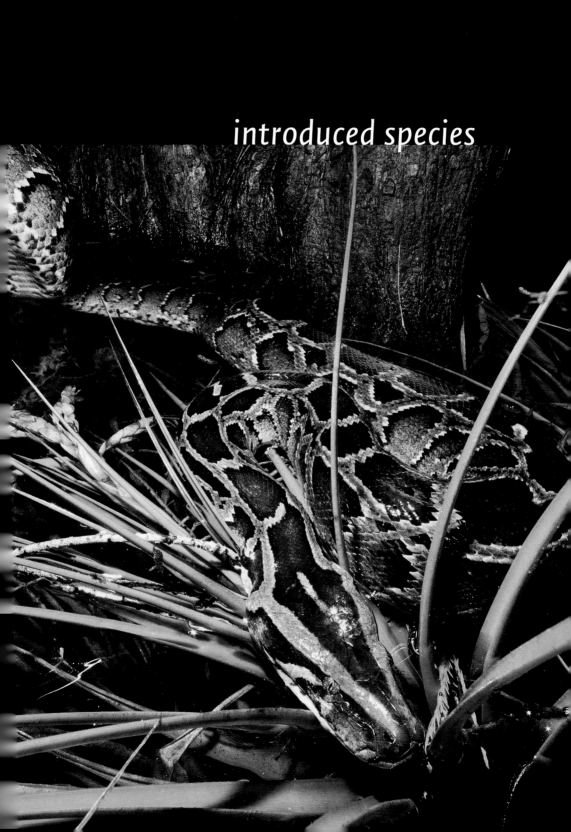

introduced species

previous page Burmese pythons are one of the many species of exotic reptiles now established in southern Florida.

For centuries, humans have transported animals and plants around the world—sometimes deliberately and sometimes by accident. As a consequence, many species are now found in areas where they do not naturally occur. Most introduced species do not survive. Some, however, adapt particularly well to their new environment and may be extremely successful, in some cases more so than native species. Such introduced species can cause many problems both for humans and for local animal species. Notable examples of introduced species considered environmentally detrimental in the eastern United States are kudzu and fire ants.

Many species of reptiles have been introduced into nonnative regions throughout the world. Mediterranean geckos, which have been widely introduced in warm regions of the United States, have caused few or no apparent problems. But other introduced species have been extremely harmful to local species.

Snakes have several characteristics that make them potentially extremely detrimental as invasive species. They are fundamentally different from some types of terrestrial apex predators such as mammalian carnivores and birds of prey. Most notably, the energetic requirements of snakes are very low, and they use the food they do obtain very efficiently. Unlike birds and mammals, they do not have to use large amounts of energy to maintain warm body temperatures and thus can convert most of what they eat into growth and reproduction. If there is plenty of food around, a snake can eat frequently and grow and reproduce rapidly. If food is limited, many snakes can easily go months and even longer than a year with no food. Snakes are also nonmigratory and nonterritorial, allowing

right Cane toads, native to tropical America, pose a danger to pets that might eat them.

below Mediterranean geckos are an introduced species that cause no apparent harm to native wildlife.

Kudzu becomes invasive in some areas, where it may cover native vegetation.

them to attain high densities in some environments. Because they are extremely efficient predators and can occur in high densities, introduced snakes can have potentially devastating effects on native prey populations.

Exotic reptiles have been and continue to be introduced in the eastern United States, most notably in Florida, both deliberately and by accident. The hospitable tropical climate of southern Florida has facilitated the establishment of numerous species of reptiles from other parts of the world. Most of the introduced reptiles are small- to medium-sized lizards such as geckos and anoles that survive well around humans and in urban environments. The caiman, a medium-sized crocodilian native to Central and South America, has been introduced in several places around Miami, and several reproducing populations have become established.

Although most of the nonnative snakes released in the United States do not survive, potentially dangerous snakes such as Egyptian cobras, king cobras, and reticulated pythons that have presumably been released by or escaped from irresponsible pet owners have been found in various places throughout the eastern United States, particularly in south Florida. The Brahminy blind snake (*Ramphotyphlops braminus*), the boa constrictor (*Boa constrictor*), the northern African python (*Python sebae*), and the Burmese python (*Python molurus bivittatus*) are all established there. The Burmese python in particular has greatly expanded its range since it was first described as a reproducing population in 2000, and it appears to pose a significant threat to native fauna and ecosystems in south Florida.

BRAHMINY BLIND SNAKE *Ramphotyphlops braminus*

SCALES
Smooth

ANAL PLATE
Several small scales

BODY SHAPE
Slender and cylindrical

BODY PATTERN AND COLOR
Shiny black or dark brown on the back; belly lighter in color

DISTINCTIVE CHARACTERS
Tiny size; belly scales same size as scales on back; tiny spine on tip of tail

SIZE
baby
typical
maximum
0"   4"   8"

Description Brahminy blind snakes are the smallest snakes in the United States, typically reaching only 3–5 inches in length. They are relatively thin and cylindrically shaped and are often mistaken for earthworms. Their snout is rounded, and their eyes are barely visible under their head scales. They are normally shiny black or dark brown above with a tan to yellowish belly. They do not have enlarged belly scales like all other snakes in the eastern United States do. The tail is short and has a small dull spine on the tip.

A rounded snout and tiny eyes are characteristic of Brahminy blind snakes.

top A Brahminy blind snake from Highlands County, Florida

All Brahminy blind snakes are females that can reproduce without being fertilized by a male.

What Do the Babies Look Like?
Baby Brahminy blind snakes look like smaller versions of the adults.

Distribution and Habitat
Apparently native to Southeast Asia and Africa, Brahminy blind snakes are now arguably the most widespread snakes in the world. They are typically associated with urbanized and agricultural areas and are found throughout much of peninsular Florida, the Florida Keys, and parts of the Florida panhandle and may even be established in southern Georgia. They have also been reported from other eastern states as well, including Alabama, Louisiana, Virginia, and Massachusetts. Brahminy blind snakes are found in moist habitats and spend virtually all their time underground or under objects such as rocks, logs, or even trash. They are often associated with ant or termite nests.

Behavior and Activity
Little is known about the activity of these snakes because of their secretive underground habits. They may come to the surface at night or during heavy rains. They are excellent burrowers that can dig headfirst into soil and disappear rather quickly. The tail spine may be used as an anchor when burrowing.

Food and Feeding
Brahminy blind snakes feed primarily on ant and termite eggs and pupae but may eat other small invertebrates.

Reproduction
Brahminy blind snakes appear to be entirely parthenogenic (i.e., populations consist entirely of females), and the females are able to reproduce without requiring sperm to fertilize their eggs. This reproductive

Brahminy blind snakes are the smallest snakes in North America and probably the most widely distributed species of snake in the world.

BRAHMINY BLIND SNAKE *Ramphotyphlops braminus*

capability likely facilitates their spread geographically, because a single snake transported to a new area can start a new population. Additionally, all individuals can produce offspring, thus increasing the number of young per generation. Three to eight small eggs are laid, likely in late spring or early summer. Juveniles are tiny, approximately 2–3 inches long.

Predators and Defense Other snakes, frogs, small birds, and mammals are either documented or probable predators. Even large invertebrates such as centipedes likely prey on these minuscule snakes. Brahminy blind snakes never

> **DID YOU KNOW?**
>
> The smallest (Brahminy blind snake) and largest (Burmese python and northern African python) snakes in the eastern United States are exotic species that are now established residents of Florida.

bite when handled but may squirm rather wildly in the hand, excrete a foul musk, and lightly poke the skin with their tail spine.

Impacts and Other Information Brahminy blind snakes are spread via the horticultural industry, especially through the transport of ornamental plants. Although they can be very abundant where they occur, Brahminy blind snakes' impacts on native ecosystems appear to be minimal. Their biology is not yet completely understood. One herpetological study reported that males of the species do exist, whereas others have suggested that the species' reproduction and genetics is very complex and that more than one species may be involved. Further research is needed to fully unravel the biology of these interesting snakes that appear to have become permanent residents in parts of the eastern United States.

What's in a Name? The species was first described in 1803 as *Eryx braminus* by François-Marie Daudin, a young French naturalist. The implied locality of the described snake was the western Bengal region of India. The genus name *Ramphotyphlops* is derived from Greek *typhlops*, meaning "blind," and *ops*, meaning "eye," a reference to the barely visible eyes, but the origin of *rampho* is unknown. The species name is a modification of "Brahmin," referring to the presumed origin of the species in India. In 2014 S. Blair Hedges and other authorities proposed that the proper scientific name should be *Indotyphlops braminus*. The uncertainty surrounding the taxonomic identity of the Brahminy blind snake is underscored by the fact that since first described more than 200 years ago it has been given 11 different generic names and 18 different species names, averaging a name change every 12 years. Brahminy blind snakes are often called flowerpot snakes because they were apparently introduced into the country in shipments of ornamental plants. They are sometimes called blind snakes or worm snakes.

BURMESE PYTHON *Python molurus bivittatus*

SCALES
Smooth

ANAL PLATE
Single

BODY SHAPE
Very stout

BODY PATTERN AND COLOR
Back light brown to gold with large, dark brown blotches; belly usually cream or yellowish

DISTINCTIVE CHARACTERS
Head distinct from neck, with dark arrow-shaped marking on top; pits visible on lips; markings on back and sides a more symmetrical pattern than on African pythons

SIZE
baby
typical
maximum
0'   10'   20'

Description Burmese pythons are among the largest snakes in the world, in some cases reaching lengths in excess of 18 feet. They are heavy bodied and have a central row of large, irregular, dark brown blotches running the entire length of the back and similar blotches on the sides. The blotches generally have a darker outline and lighter center. The background color is usually brown to golden. The head is fairly distinct from the neck and typically has an arrow-shaped marking on top. The generally unmarked belly is usually cream or yellowish. The ventral

top Burmese pythons are established in south Florida and are linked to severe declines in numerous species of once-common mammals in Everglades National Park.

BURMESE PYTHON *Python molurus bivittatus*

Large alligators are one of the few known predators of adult Burmese pythons. However, pythons themselves are able to capture and consume large birds and mammals as well as small- to medium-sized American alligators.

scales are considerably narrower than those of native snakes in the eastern United States. Several pits visible on the lips are sensitive to infrared radiation.

What Do the Babies Look Like? The babies look like miniature adults, but their colors are often more bold.

Distribution and Habitat In their native range, Burmese pythons occur through most of Southeast Asia and along the India-Nepal border. They are found in a wide variety of habitats, including tropical rainforests, grasslands, scrub-desert, temperate forests, and sometimes even brackish waters such as mangrove and salt marsh areas. They live in fairly temperate regions on the Tibetan Plateau and in the foothills of the Himalayas in Nepal. The extent of their distribution in south Florida is difficult to ascertain because of their secretive nature. They were first found in Everglades National Park near the West Lake/Flamingo area and have since spread east and west and well north of Alligator Alley (Interstate 75). They are present on Key Largo and on the west coast of the Florida peninsula in the vicinity of Naples. Burmese pythons are found in nearly every Florida habitat, both natural and altered. They are usually found in areas near fresh or brackish water and actually spend considerable time in aquatic habitats. They still appear to be abundant in Everglades National Park, although hundreds have been removed from there. Burmese pythons appear to have the ability to traverse relatively large bodies of saltwater and are still abundant in the southern mangroves of Everglades National Park where they were first found.

Baby Burmese pythons are patterned similar to adults but may appear brighter in coloration.

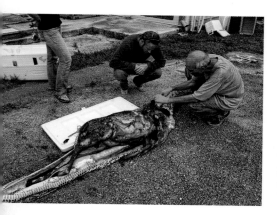

above A 15-foot-long python is dissected after eating an 80-pound white-tailed deer.

left A Burmese python swallowing a great blue heron

The majority of Burmese pythons encountered by scientists who study them are collected at night crossing roads in Everglades National Park, Florida.

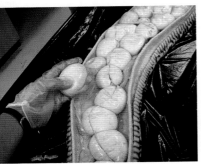

Burmese pythons average about 45 eggs per clutch and have been documented to sometimes lay more than 80 in Florida.

Behavior and Activity In Florida, Burmese pythons can be active in any season and at any time but spend most of their time hidden either in underground retreats or in water, often with only their snout protruding periodically above the surface to breathe. During warmer months they appear to be more active at night and are often found crossing roads. During winter they bask in the sun along canal banks and other open areas. Pythons that were relocated from their capture locations and released as part of a radiotracking study outside Everglades National Park returned within a few months to the areas where they were captured, even those that were displaced more than 20 miles. How they navigated such long distances to return to their home area remains unknown.

Food and Feeding Burmese pythons are large constricting snakes that probably rely primarily on ambush tactics to capture prey. Little is known about the diet in their native range, but in general they eat a wide variety of birds and mammals and, occasionally, crocodilians. In south Florida they are known to eat most native mammals other than bats, and have been documented eating rats, rabbits, opossums, raccoons, bobcats, and even fully grown white-tailed deer. They also prey on a wide variety of birds, including species that are of conservation concern such as wood storks and limpkins. They have been documented feeding on alligators on numerous occasions and have even eaten domestic cats, ducks, geese, and chickens.

Reproduction Female Burmese pythons in Florida have been documented to lay up to 87 eggs (typical clutch size is about 45), generally in late spring or early summer. The eggs hatch approximately 2 months later, and the babies are generally between 1.5 to 2 feet long. After a female Burmese python lays her eggs, she coils around them, presumably to defend them against potential nest predators. The females also exhibit "shivering thermogenesis," a behavior that raises their body temperature substantially above ambient temperature (i.e., they become warm-blooded or endothermic) and provides a warmer, more stable environment for egg development.

Predators and Defense Little is known about predators of Burmese pythons in their native range. In

This 13-foot Burmese python swallowed a 6-foot-long alligator and then died and burst open in the heat of Everglades National Park.

The safe capture and handling of large pythons in Everglades National Park generally requires more than one experienced individual.

Burmese Python 353

A Burmese python in the Everglades

Florida, alligators feed on juveniles and adults and may represent a major source of mortality for younger individuals. Large birds such as hawks and herons, indigo snakes, and a variety of mammals likely feed on young Burmese pythons. As they grow, the number of potential predators decreases, and large alligators are probably the primary predators of large adults. When threatened, Burmese pythons will generally try to escape, but if grabbed or harassed they will strike, hiss, and attempt to bite repeatedly. Although nonvenomous, a large python can inflict large, painful wounds. Pythons will also thrash wildly if grabbed and will smear their attacker with feces and musk.

Impacts and Other Information In 2012 scientists determined that Burmese pythons were the likely cause of severe declines in several species of once-common mammals in Everglades National Park. They measured the relative abundances of deer, bobcats, foxes, rabbits, raccoons, and opossums before and after Burmese pythons began proliferating (approximately 2000 or 2001). Within only a few years after python numbers increased, the relative abundances of raccoons and opossums dropped 99 percent, deer 94 percent, and bobcats 87 percent. Foxes and rabbits are no longer found in the park, although rabbits had once been abundant. The declines appear to be directly attributable to predation by Burmese pythons. The impacts pythons are having on other species that are less easy to monitor have not been determined.

Efforts to develop methods to control Burmese pythons in Florida have thus far been unsuccessful. Development of effective management tools is extremely difficult because of the snakes' secretive nature, the large geographic area over which they occur, and the relative inaccessibility of many sites. At this time there is no known method that shows promise of suppressing Burmese python populations across any substantial portion of their range in Florida. How far and how quickly they will spread northward is unknown, but several climate models project suitable climate for this species through much of the southeastern United States.

What's in a Name? Python molurus was described by Carl Linnaeus in 1758, and Heinrich Kuhl described Python bivittatus in 1820. Taxonomists continue to disagree regarding the species status of Burmese pythons (Python molurus bivittatus). Most scientists agree that the Burmese python should be considered a subspecies of the Indian python (P. molurus), although others contend that it should be considered a distinct species, P. bivittatus, on the basis of small differences in scale and color patterns. In Greek mythology Python was a giant snake (actually some paintings show it to be a winged dragon) slain by Apollo, making it an apt choice for the genus name. The species name is derived from molouros, meaning "a type of snake" in Greek, and the Latin bi, meaning "two," and vitta, meaning "stripe," presumably referring to the two stripes on the head. Burmese pythons are sometimes called Asian pythons or Asian rock pythons. In Florida, most people just call them pythons.

Burmese pythons have been documented to reach lengths of more than 18 feet. This large specimen is being taken for examination by National Park Service biologist Skip Snow.

NORTHERN AFRICAN PYTHON *Python sebae*

SCALES
Smooth

ANAL PLATE
Single

BODY SHAPE
Very stout

BODY PATTERN AND COLOR
Back yellowish brown to gold with connecting dark brownish blotches down center; sides with dark-bordered blotches; belly white to cream

DISTINCTIVE CHARACTERS
Arrow shape on head bordered by light lines; pits on upper lip; markings on back irregular

SIZE
baby
typical
maximum
0' 10' 20'

Description Northern African pythons are closely related to the much more widespread Burmese pythons, grow to a similar size (approaching 20 feet), and have a somewhat similar color pattern. The dark blotches on the back of northern African pythons, however, connect in irregular ways, forming a pattern that often results in irregular stripes along the back. The background color is generally yellowish brown or gold. Dark-edged blotches are present along the sides, and the belly is white or cream colored.

top The markings and pattern on the back of a northern African python are more irregular than the Burmese python.

NORTHERN AFRICAN PYTHON *Python sebae*

Like its close relative the Burmese python, the northern African python has dark lines bordered by light lines that converge to form an arrow shape on the head.

The small number of northern African pythons found in Florida have been restricted to a region west of Miami.

What Do the Babies Look Like? The babies look like miniature adults but often have a slightly more contrasting pattern.

Distribution and Habitat In their native range, northern African pythons are found throughout western and central sub-Saharan Africa, where they occupy a variety of habitats. In Florida they have been found only in a relatively small area outside the northeast corner of Everglades National Park. This area of canals, swampy areas, and ponds interspersed with forested and brushy habitat is heavily modified by human activity. This species is probably capable of inhabiting nearly any habitat in south Florida.

Behavior and Activity Little is known about the behavior of this species in the native range or in Florida. During cool weather they are likely to bask in the sun, but they are presumably nocturnal during warmer weather and likely tend to prefer areas near water.

Food and Feeding Like all pythons, this snake kills its prey by constriction and likely hunts primarily using ambush tactics. Limited diet records are available for northern African pythons in Florida, but like Burmese pythons, they can eat nearly any bird or mammal found in their habitat.

Reproduction The females lay a large number of eggs in the late spring or summer and then coil around them during incubation. In Florida, nests with more than 20 hatched eggs have been found. Larger females probably lay larger clutches. The babies are approximately 1.5–2 feet long when they hatch.

Predators and Defense Little is known about predators of this species in Florida, but a number of large birds, alligators, and midsize to large mammals likely prey on smaller individuals. As they grow, fewer and fewer animals pose a risk, and large alligators and humans are probably the only species that pose any real danger to large adults. When threatened or harassed, they will strike and bite, and like Burmese pythons exude musk and fecal material on their attacker.

Impacts and Other Information Compared with Burmese pythons, only a small number of northern African pythons have been found in Florida—all restricted to a relatively small geographic region slightly west of Miami. Whether they will expand their range and increase in numbers the way Burmese pythons have done is unknown. In captivity they tend to be less docile than Burmese pythons and are thus less common in the pet trade. Their irascible nature in captivity has led to speculation that they will become more of a threat than even Burmese pythons or that they will hybridize with Burmese pythons and produce a "super snake." The reality is that their defensive behavior toward humans in captivity likely has nothing to do with their potential impacts on natural ecosystems. Also, if the two python species do produce hybrids, these will probably be similar in size and intermediate in appearance to their parents and no more of a substantial threat than these snakes already are.

What's in a Name? Johann Friedrich Gmelin described *Coluber Sebae* in Carl Linnaeus's *Systema Naturae* in 1789, one year after Linnaeus died. The species name honors Albertus Seba (which explains why Gmelin capitalized the species name in the original description), a Dutch pharmacist who collected exotic animals that sailors brought to the harbor at Amsterdam. Linnaeus obtained many original species descriptions from Seba. Northern African pythons are also called African pythons, northern African rock pythons, African rock pythons, or just rock pythons.

BOA CONSTRICTOR *Boa constrictor*

SCALES
Smooth

ANAL PLATE
Single

BODY SHAPE
Very stout

BODY PATTERN AND COLOR
Body tan or gray with dark blotches; tail yellowish with reddish blotches; belly lighter with darker blotches

DISTINCTIVE CHARACTERS
Thick, triangular head distinct from neck; broad dark line running from behind each eye to back of jaw

SIZE
baby
typical
maximum
0'   6'   12'

Description Boa constrictors are large, heavy-bodied constricting snakes that are generally tan or gray with a series of interconnecting saddles or bars, each of which usually contains a white spot. Corresponding dark blotches along the sides generally have a light center. The tail differs in color from the rest of the body, with the background usually becoming more yellowish. Saddles on the tail become more blotchlike and are darker and more reddish. The belly is cream or gray with irregular darker spots. The head is triangular and well distinguished from the neck.

What Do the Babies Look Like? The babies look like miniature adults.

top Boa constrictors have robust bodies that are tan or gray with dark saddles across the back.

Boa constrictors have a thick, triangular head distinct from the neck and a broad dark line running from behind each eye to the back of the jaw.

BOA CONSTRICTOR *Boa constrictor*

A boa constrictor from Dade County near the city of Miami

Distribution and Habitat Boa constrictors range naturally from northern Mexico southward through Central America to northern Argentina. They have been introduced onto the islands of Cozumel, Aruba, and Puerto Rico in addition to southern Florida, where a breeding population exists in a Miami-Dade County park, the Deering Estate at Cutler on Biscayne Bay. More than 100 boa constrictors have been found in the park since the population was discovered in 1989. Several individuals have been found on the Florida Keys, and in other areas as well, but no other known breeding population exists in the United States. In their native range they occupy a wide variety of habitats, ranging from dry desert scrub to tropical rainforest. The subtropical climate has undoubtedly facilitated their success in south Florida.

Behavior and Activity Boa constrictors are found over a large geographic area in their native range, and the particular habitat and climate in which they occur influence their activity and behavior. They may be active at night or day, but little else is known about their activity in Florida. During hot weather they are probably crepuscular or nocturnal, but during cooler weather they may be more active during the day. They seek shelter in stump holes, gopher tortoise burrows, and tree holes.

Food and Feeding In their native range, boa constrictors generally feed on mammals, birds, and large lizards that they kill by constriction. In Florida they have been documented preying on opossums, rabbits, rats, mice, round-tailed muskrats, and squirrels. Birds have also been documented in their diet, including grebes, ibises, and a limpkin. One boa constrictor was confirmed to have killed and eaten a domestic cat.

Reproduction Boa constrictors give birth to live young, and litters are reported to range from 2 to 64 babies. Litters of the few gravid females found in Florida ranged in size from 22 to 47. Females in Florida appear to give birth in mid- to-late summer.

Predators and Defense Large snake-eating snakes, bobcats, foxes, and large birds presumably prey on young, small boa constrictors. Adults have fewer predators as they become larger and more formidable. When threatened, boa constrictors will strike at and bite an aggressor. If grabbed, they may coil around their attacker and exude a noxious musk.

Impacts and Other Information Boa constrictors in Florida do not appear to be expanding their range appreciably, and thus the overall impacts on ecosystems in the state or elsewhere are likely minimal. No study has documented any appreciable impacts on native prey populations or domestic animals, although it is certainly possible that boa constrictors have reduced prey populations in the areas where they occur.

What's in a Name? Carl Linnaeus described this species in 1758 stating "Habitat in Indiis," referring to the West Indies. *Boa* means "large snake" in Latin. The boa constrictor is one of the few North American species of snakes that has retained the name originally given it by Linnaeus and is the only one whose scientific name and common name are the same. For many years in the 1900s *Constrictor* was the accepted genus name, and a popular pet trade subspecies known as the red-tailed boa thus had the scientific name *Constrictor constrictor constrictor*. Boa constrictors are often just called boas, but several other species also belong to the boa family, so it is best to be specific.

people and snakes

WHAT IS A HERPETOLOGIST?

Herpetologists are scientists who study reptiles and amphibians (also known collectively as herpetofauna). Most herpetologists specialize in a particular discipline or disciplines of biology, such as ecology, physiology, or genetics, and their research often has important implications for other groups of animals in addition to reptiles and amphibians. Many herpetologists today focus on conservation issues related to reptiles and amphibians—for example, investigating the impacts of forestry practices on woodland salamanders or how some reptiles can survive in urbanized environments while others cannot. Others study the effects of pesticides on reptiles and amphibians, many of which are particularly sensitive to these chemicals. Countless other research projects involve herpetofauna—so many that some scientific journals are devoted entirely to publishing research about amphibians and reptiles.

Many, perhaps most, professional herpetologists experienced an early taste of their future careers during childhood as they pursued an interest in snakes by collecting them and keeping them as pets. Although some herpetologists devote their careers to studying other groups of herpetofauna (frogs, salamanders, lizards, turtles, or crocodilians), the primary appeal for many continues to be snakes. It is a natural extension of an exhilarating journey from youth into adulthood, a journey that for most will last a lifetime.

Many herpetologists are fortunate enough to pursue a career that allows them to have as much fun as adults as they did when they were children.

X-radiography can be used to identify objects inside a snake, including radiotransmitters, PIT tags, and bones of prey. This large ratsnake has eaten nine wood duck eggs.

Why Do Herpetologists Study Snakes? Herpetologists study snakes for a variety of reasons. While sheer fascination with a particular species or with snakes in general has stimulated many a career in herpetology, that alone does not justify spending considerable money, time, and effort getting to know them better. Some snakes are excellent model organisms for studies of important biological phenomena. For example, research on snakes' metabolism and their ability to fast for long periods might provide insights related to the dietary and digestive issues of domestic animals and humans. In the case of endangered snake species, studies of their basic ecology and behaviors are necessary to understand how to preserve them and their habitats. Other species of snakes, especially some aquatic species, may be good indicators of environmental quality of a habitat. Studies evaluating the status of these animals may indicate potential environmental problems.

How Do Herpetologists Study Snakes? Herpetologists use a variety of techniques to study snakes. Because snakes are often very secretive, capturing them may require special methods. Generally, the best way to catch snakes is to spend a lot of time in their natural habitats. Watersnakes can be grabbed from trees while they are basking over the water, and many snakes can be found crossing roads.

One effective way to capture snakes systematically in an area where they occur is to use a drift fence, which is simply a "wall" made of silt fencing or aluminum flashing with funnel traps or buckets placed alongside it at intervals. Snakes that encounter the drift fence tend to turn and follow it, and eventually enter or fall into the traps. Another very effective method for capturing snakes is to use large flat objects, or "coverboards"—usually sheets of plywood or tin—that are placed on the ground and later lifted to see what has taken refuge beneath them.

Captured snakes are often measured and marked for future identification by herpetologists; some researchers take tissue samples for DNA analysis. Some snakes can be easily marked by clipping small

A funnel trap placed alongside a log, building, or drift fence can be used to capture both large and small snakes.

Looking beneath coverboards is one of the simplest techniques for finding snakes for both professional and amateur herpetologists. In regions with large venomous species it is important to stand several inches away from the edge of the coverboard when it is lifted.

PIT tags that can be checked with a handheld reader have become a commonly used technique for permanently identifying individual snakes.

sections from their belly scales in identifiable patterns. A more high-tech method involves the use of passive integrated transponders, or PIT tags (RFID tags), which are harmlessly injected into the snake's body cavity and then later read by a device similar to a bar code reader in the grocery store. Each PIT tag has a number that individually identifies a particular snake.

Many herpetologists use radiotelemetry to study snakes' natural behaviors in the field. A radiotransmitter in the snake emits a signal at a particular radio frequency that allows a researcher with a receiver and a special antenna to follow it. Some special transmitters even have sensors that allow the researcher to monitor the snake's body temperature. Because a snake cannot wear a radio collar like most other wild animals that scientists study with radiotelemetry, transmitters typically are surgically implanted into the snake's body cavity. This simple surgery causes no harm to the snake if done correctly by an experienced herpetologist or veterinarian.

What Rules Must Herpetologists Follow to Study Snakes? Herpetologists who collect snakes for research or educational purposes, such as during a college class in herpetology, must know and follow the state and federal laws that protect snakes and other wildlife in their area. Most, if not all, states require anyone who wishes to remove a snake from the wild for research purposes to have a scientific collecting permit, and some states place limits on how many individuals of certain species can be kept at one time. Snakes in the eastern United States that are protected under the Endangered Species Act (such as indigo snakes and copperbelly watersnakes) can be handled in the field or kept in captivity only by someone with a special federal permit. Although learning the laws, which differ from state to state, and obtaining permits may require paperwork and payment of a fee, most rules and regulations are intended for the protection and overall best interest of the animals.

Herpetologists conducting research on snake movement patterns use radiotelemetry to determine the locations of individuals with surgically implanted transmitters.

Only people with expertise in safely handling venomous snakes should attempt to capture them. Here, an experienced herpetologist uses a simple and safe technique for holding a large eastern diamondback rattlesnake. Keeping the snake's head and front part of its body in the tube is safe for the handler and for the snake.

URBAN SNAKES

Although many snakes require relatively natural environments, several kinds of snakes are commonly found in areas where people live. These "urban snakes" offer many opportunities for people to become familiar with a fascinating element of nature and to educate others about them as well. Snakes were on this planet millions of years before humans and houses arrived, so a backyard or a city park cannot be viewed as the natural habitat of any snake. However, gartersnakes, ratsnakes, and even venomous copperheads can be relatively common in some urbanized areas. At least one species common throughout much of the eastern United States, the brownsnake, is found more often in suburban and urban areas than in its natural habitat in the woods!

An individual of any species can potentially turn up in someone's yard because people have invaded every natural habitat in the eastern United States to some degree. Wild animals that have been

left Kirtland's snake can be found in vacant lots in most larger cities within its geographic range.

below A common gartersnake could possibly show up in a residential neighborhood in any eastern state.

displaced from their homes may roam in search of food or shelter and may unwittingly enter a newly developed housing area or even a shopping center. As time goes by, most wildlife disappears, including snakes. But some species persist.

Several small, secretive species of snakes do well around houses as long as adequate places to hide are present and their prey base of earthworms, slugs, and insects has not been damaged with pesticides. Ring-necked snakes and the earthsnakes can turn up under rocks, logs, or other hiding places in a backyard. Although less secretive than the smaller species, the common gartersnake can show up almost anywhere, too, including in someone's yard or on their porch. Rough green snakes often do quite well in neighborhoods with shrubs, vines, and bushes, but their presence often goes unnoticed because of their excellent camouflage. One of the most unexpected of the backyard snakes is the scarlet snake, sometimes referred to as the "swimming pool snake" in sandhill-area communities because individuals frequently fall into swimming pools during their nighttime activities.

> **DID YOU KNOW?**
>
> No chemical will effectively repel all snakes without having a major effect on other wildlife and domestic pets as well. Some people claim that sulfur or mothballs will deter snakes. However, you would probably have to use so much sulfur to keep the snakes out that it would keep you from entering the house as well.

In areas with lakes, canals, or even retention ponds, northern watersnakes and banded watersnakes often thrive and become part of the common fauna. Backyard water features, especially koi or goldfish ponds, often attract watersnakes. Queen snakes have been known to persist in developed areas where streams occur as long as conditions remain relatively unpolluted and crayfish are present.

Less welcome, yet sometimes common, residents of suburban areas are two venomous species—copperheads and pigmy rattlesnakes. When present in an area, both species are probably much more common than the human inhabitants realize because the snakes spend much of their time hiding or inactive. They are seen most often during the spring and fall as they move to and from hibernation sites or in search of mates or prey.

Ratsnakes are the most likely species to be found inside a house in many areas of the eastern United States because of their climbing ability and relative abundance in many urban and suburban areas. Additionally, the rodents often found around urbanized areas may attract ratsnakes. However, any snake can

Most Common Snakes Found in 25 of the Largest Eastern Cities

Species given for each city are based on records by one or more local professional herpetologists in the area who have responded to identification queries from the general public or examined records of the most commonly reported species of snakes. The association of some snakes with a city may be very localized due to particular habitat conditions in a specific part of a city and does not necessarily mean that the species occurrence is widespread in the metropolitan area.

Species	25 cities	Atlanta	Baltimore	Boston	Charlotte	Chicago	Cincinnati	Cleveland	Columbus	Detroit	Indianapolis	Jacksonville	Louisville	Memphis	Miami	Milwaukee	Nashville	New Orleans	New York	Orlando	Philadelphia	Pittsburgh	Richmond	Tampa	Toledo	DC
Thamnophis sirtalis Common Gartersnake	24	■	■	■	■	■	■	■	■	■	■	■	■	■		■	■	■	■	■	■	■	■	■	■	■
Storeria dekayi Brownsnake	19	■	■	■		■	■	■	■	■	■	■	■	■		■	■		■		■	■	■		■	■
Nerodia sipedon Northern Watersnake	18	■	■	■		■	■	■	■	■	■		■	■		■	■		■		■	■	■		■	■
Pantherophis obsoletus Ratsnake	14	■	■						■			■	■	■			■		■	■		■	■	■	■	
Coluber constrictor Racer	13	■		■					■			■	■	■			■		■	■			■	■		
Diadophis punctatus Ring-Necked Snake	10			■			■		■	■			■	■				■	■			■				
Lampropeltis triangulum Milksnake	9	■		■									■			■			■			■			■	■
Storeria occipitomaculata Red-Bellied Snake	5			■					■							■						■			■	
Nerodia fasciata Southern Banded Watersnake	4											■		■	■			■		■						
Lampropeltis getula Common Kingsnake	4	■										■								■				■		
Pantherophis guttatus Corn Snake	3											■			■					■						
Thamnophis sauritus Eastern Ribbonsnake	3										■				■					■						
Micrurus fulvius Coral Snake	3											■			■					■						
Opheodrys aestivus Rough Green Snake	3											■											■	■		
Agkistrodon contortrix Copperhead	2	■																								
Clonophis kirtlandii Kirtland's Snake	3						■				■		■													

Species	# in cities	8	5	7	4	5	7	6	13	9	6	11	6	11	3	4	5	7	7	4
Ramphotyphlops braminus Brahminy Blind Snake	3								■			■							■	
Thamnophis butleri Butler's Gartersnake	2								■					■						
Tantilla relicta Florida Crowned Snake	2							■												
Pantherophis vulpinus Fox Snake	2					■													■	
Agkistrodon piscivorus Cottonmouth	2								■											
Sistrurus miliarius Pigmy Rattlesnake	1								■											
Carphophis amoenus Eastern Wormsnake	2									■		■								
Nerodia rhombifer Diamondback Watersnake	2									■		■								
Thamnophis radix Plains Gartersnake	1												■							
Virginia valeriae Smooth Earthsnake	1	■																		
Haldea striatula Rough Earthsnake	1									■										
Heterodon platirhinos Eastern Hognose Snake	1											■								
Thamnophis proximus Western Ribbonsnake	1											■								
Crotalus horridus Canebrake/Timber Rattlesnake	1											■								
Nerodia cyclopion Western Green Watersnake	1									■										
Regina septemvittata Queen Snake	1											■								
Python molurus bivittatus Burmese Python	1										■									
Nerodia erythrogaster Plain-bellied Watersnake	1						■													

species in city

readily enter a house with doors that are at ground level, and domestic cats often bring live snakes and other animals into houses. Once a snake is removed from a house, it is unlikely to be seen again.

Some species of eastern snakes, especially the large rattlesnakes, almost never persist in developed areas. Unless extensive hiding sites are available, individuals are likely to be discovered and are unlikely to survive most encounters with people. Additionally, the chances of road mortality are extremely high for such slow-moving snakes.

What do the species of urban snakes have in common? No general rule can explain why some species persist, some thrive, and others disappear completely from habitats when humans move in. The important point is that some snakes do remain, and their survival encourages us to understand that we can share our surroundings with them and to appreciate them as important parts of our urbanized ecosystems.

top left One of the most commonly encountered species to persist in city parks, backyards, and vacant lots in many urban areas is the brownsnake, a species found in every eastern state. Brownsnakes are seldom more than a foot long and are completely harmless. They rarely if ever bite people, even when handled. These two are from Champaign County, Ohio.

below Northern and banded watersnakes are common visitors to backyard water features throughout the eastern United States. Both species eat fish and amphibians. Ornamental fish such as koi, as well as frogs and tadpoles that use the water for breeding pools, provide a ready source of food for snakes that travel from nearby wetlands.

Ratsnakes are the most frequently encountered large snake to enter houses or other structures in cities and suburban areas throughout much of the eastern United States. Ratsnakes have a remarkable ability to climb up trees and brick or wooden walls in search of bird or squirrel nests.

More and more people are becoming educated about and interested in the wildlife around them, including snakes, and appreciate finding an occasional nonvenomous species of snake. Homeowners can take several measures to encourage snakes to take up residence in their backyards. Nearly all snakes want places where they can hide. Boards, pieces of tin, and rock piles are all places where snakes feel safe because they can hide from view. Vegetation that provides hiding places, such as dense shrubbery or tall grass, is likely to attract more snakes than carefully trimmed lawns. Nearby water, such as a pond or small stream, may add to the likelihood of encountering a snake in a yard because almost all aquatic snake species spend some time on land. Although a yard with numerous hiding areas may be "snake friendly," unless snakes are already in the area it may still be devoid of snakes. In addition to hiding places, snakes require different forms of prey. One means to ensure suitable prey for many small snakes is to be cautious about applying pesticides that can accumulate in the animals that snakes eat, such as earthworms, slugs, and insects.

Some people still prefer to keep all snakes away from their gardens, yards, and houses. The bad news for those people is that there is no effective snake repellent. Mixtures of sulfur and naphthalene (mothballs) are sold by some companies with the promise that they will get rid of snakes, but no product has

Ring-necked snakes are one of the species of harmless snakes often found in suburban neighborhoods in the eastern United States.

been proven effective for this purpose except in laboratory tests conducted in enclosed systems. In most situations, the application of enough of any chemical to deter snakes would harm many other animals and would likely also be a nuisance to the human inhabitants. Barriers such as silt fencing or vertical aluminum flashing may effectively deter snakes in some situations, but many snakes are good climbers, and others may crawl under barriers that are not embedded deeply in the soil.

Dealing with venomous snakes around a house or the possibility of their presence can pose a dilemma when small children or outdoor pets are present, even for people who appreciate snakes. Removing a venomous snake from an area inhabited by people is best done by a local snake expert from a university, nature park, or environmental education center. You may also be able to find herpetologists at a regional reptile and amphibian society who have experience handling venomous snakes.

SNAKES AS PETS

Many people find snakes fascinating, and snakes are often kept as pets. Captive snakes, unlike more commonly kept pets such as dogs, cats, and birds, generally eat infrequently, produce little waste, generally do not carry diseases transmittable to humans when properly cared for, do not require a lot of room, and do

A well-constructed cage with a hiding place and water bowl is essential for proper care of a captive snake.

When properly cared for, common kingsnakes make excellent pets for children.

not make noise. Therefore, as long as you understand the particular snake's requirements, many are fairly easy and safe to maintain in captivity.

If you are interested in obtaining a snake to keep at home, animals bred in captivity generally make far better pets than wild-caught animals. Many reptile breeders produce a wide variety of high-quality captive-bred snakes that sell for relatively low prices compared with those charged by pet stores. Search for them on the Internet. The most preferred captive-bred animals are ones born from parents that were also born and raised in captivity. Captive breeding produces snakes that are healthier than many wild-caught individuals and also provides an environmental service by helping to reduce the removal of snakes from the wild. Among species native to the eastern United States, several varieties of kingsnakes and corn snakes are very colorful and make excellent pets.

Before purchasing any snake, you should first check local, state, and federal laws that may restrict your ability to legally own exotic or native animals. Some states, for example, have very specific regulations restricting the release of pet snakes into the wild. Then make sure you understand the snake's requirements in captivity. Many books are available on the subject, and much useful informa-

> **DID YOU KNOW?**
>
> Although albino snakes are commonly bred in captivity, they are extremely rare in the wild.

tion can be found on the Internet as well. Venomous snakes and very large boas or pythons can potentially be dangerous and should not be kept as house pets under most circumstances.

Captive snakes require a suitable cage. Snakes are remarkable escape artists, and they can be very difficult to find once they are loose in most people's houses, so make sure that the cage you choose is secure. Aquariums work well for many species because they are inexpensive and easily cleaned, and lids can be purchased that fit them well. Newspaper, paper towels, or aspen bedding provides a suitable substrate for most species. A water dish filled with clean water must be available at all times. Most snakes will use a "hide box" placed inside the cage and should at least be given a folded newspaper or other material in which they can hide. Snakes must also be kept at appropriate temperatures. Special heaters can be purchased that heat one end of a cage and create a gradient, allowing the snake to choose from a range of temperatures. Kingsnakes, ratsnakes, and many other species will usually feed eagerly on either freshly killed mice available from a pet store or frozen mice that have been thawed.

If you decide you want to keep a snake as a pet, make sure that you are motivated by a sincere interest in the animal and not just fascinated by the idea

Events such as the Repticon Reptile and Exotic Animal Conventions that are open to the public in many eastern cities give children and adults opportunities to get an up-close view of a variety of snake species they might not otherwise see. Snakes sold as pets at such events are typically captive bred and in healthy condition.

380 People and Snakes

Despite their gentle disposition, southern hognose snakes do not make good pets because they require a specialized diet of toads and may be difficult to keep healthy in captivity. Additionally, because they are protected throughout much of their range, it may be illegal for a person to keep one in some states.

of keeping something of which many people are afraid. In fact, you should never deliberately scare anyone with a pet snake. Doing so enhances people's negative attitudes toward snakes and toward those who have an interest in their welfare. Additionally, if you are motivated by the "thrill" of keeping a snake, that "thrill" may soon diminish, often resulting in neglect of the animal. Finally, wild-captured snakes should be released only at the site where they were captured, and captive-bred snakes should never be released into the wild.

A strikingly beautiful scarlet kingsnake can sometimes convince people who are wary of all snakes that they are not necessarily bad.

SNAKE CONSERVATION

Although they are feared or even despised by many people, snakes deserve the same consideration given to any other animal. Snakes are valuable elements of our natural ecosystems. They serve both as important prey for many animals and as predators that help control other animal populations. In some areas they are the top predators. Some snakes can be found in extremely high densities and thus form a large portion of the animal biomass and "stored energy" in the ecosystems in which they live. Watersnakes in some areas can be extremely abundant and serve as important prey for many animals, including birds such as herons and egrets. Ratsnakes are also abundant in many habitats, especially around farms, where they play an important role in helping to control rodent populations. In fact, some farmers actually release ratsnakes or corn snakes around their barns to eat rats and mice.

Snakes can also serve as important bioindicators or biomonitors of environmental integrity. Their absence or lack of abundance may signal general problems in the environment. For example, many watersnakes cannot survive in unhealthy (e.g., contaminated by pesticides) wetlands, and thus their absence can indicate underlying problems in the environment that are not otherwise ob-

Despite having an overrated reputation as being aggressive, cottonmouths bite very few people unless they are picked up, and they prefer to avoid humans whenever possible. Negative attitudes about the potential threats by U.S. venomous snakes are generally based on superstitions and misinformation that have no basis in fact.

vious. Additionally, because snakes are predators and are high in the food chain, some toxins tend to accumulate in their tissues and thus may be detectable in snakes while being very difficult to detect in the environment.

Finally, as an integral part of natural ecosystems, snakes are no less worthy of our respect and admiration than whales or eagles or other animals on which humans place a high value. Snakes add considerably to the biodiversity of the eastern United States, and our encounters with them provide many memorable experiences. How many people would ever forget the fascinating spectacle of a ratsnake crawling up the trunk of a tree or a watersnake swallowing a catfish?

> **DID YOU KNOW?**
>
> Snakes do not chase people. If a snake moves toward a person, it probably just happens to be moving in that direction; or it may be trying to get to a hole or some other hiding place, and the person is in the way.

Threats to Snakes Unfortunately, like other wild animals, and far more so than most, snakes suffer as a result of the activities of humans. Habitat destruction

Road building and commercial development in natural habitats are among the greatest environmental threats to many species of snakes.

and alteration is by far the primary conservation concern for snakes in the eastern United States. Although a few species of snakes may persist around some human habitations, most die out whenever their habitat is altered significantly or destroyed. Human development not only destroys snake habitats, but roads, parking lots, and commercial structures separate the remaining pieces of habitat, isolating populations of snakes and making them more prone to extinction. Mortality on roads is a continual threat to snakes during warm months in all parts of the eastern United States, and intentional killing of all snakes continues in areas where a sense of environmental stewardship has not become part of the culture among the residents.

Although populations of most snake species have declined throughout the eastern United States because of human activities, some species have been harmed more than others. Indigo snakes and southern hognose snakes have disappeared from much of their previous ranges as a result of human activities. The copperbelly watersnake is endangered and listed as threatened by the Endangered Species Act because of habitat destruction and fragmentation of the wetlands it requires. The eastern diamondback rattlesnake, the largest species of rattlesnake in the world, is now apparently extinct in Louisiana, and only remnants of populations persist in North Carolina. Likewise, southern hognose snakes can no longer be found in Alabama or Mississippi. Clearly, the decline of snake populations in much of the eastern United States reflects the situation of many wildlife species whose survival is not compatible with widespread habitat loss and degradation.

Conservation Laws Persuading legislators to pass laws protecting snakes, especially venomous species, is a difficult task. Laws and regulations designed to protect certain species from commercial collection or wanton killing have been implemented in some states. Few if any laws effectively prevent habitat from being destroyed solely because an area serves as an important environ-

Some species of snakes, such as the glossy crayfish snake, are so seldom encountered, even by herpetologists, that their habitat and food source (crayfish) can be destroyed by commercial development or pollution before conservation biologists are even aware of the presence of glossy crayfish snakes in an area.

Because they are so small (note pinecone in background) and well camouflaged, pigmy rattlesnakes are often more common than is apparent. Consequently, entire populations can be unintentionally eliminated by commercial development or timber operations that till the soil and remove vegetation.

A canebrake rattlesnake takes a major risk as it enters a highway in southern Georgia. Roads are responsible for the deaths of thousands of snakes each year, especially in the southern states.

ment for snakes. The indigo snake, copperbelly watersnake, and Atlantic salt marsh snake are among the very few snakes in the entire country to receive any federal protection under the Endangered Species Act. But laws and regulations are only as effective as those who enforce them. Limited manpower for state and local governments may result in little enforcement of such laws, and some law officers and judges remain among those who have an irrational fear and dislike of snakes and do not uphold the laws designed to protect them.

Snakes' Best Hope The best hope to protect the natural habitats of snakes and prevent malicious killing lies in changing public attitudes through education. People who understand the importance of snakes as natural parts of our ecosystems are more likely to support laws to protect them. It is important to remember that habitat loss is not a threat to snakes alone but is also the primary threat to other wildlife. Snakes will nearly always benefit from habitat protection designed for other animals, and vice versa. Once the general public

Teaching children to appreciate snakes can create positive public attitudes about reptile conservation. Children can easily learn to identify ring-necked snakes as an eastern species that is unlikely to be confused with any native venomous snake.

Educating children about reptiles is a positive step toward conservation. With proper instructions children can learn to identify and catch harmless snakes, such as this ratsnake from Sullivan's Island, South Carolina. Lessons on snake safety are critical before allowing anyone to capture and handle a wild-caught snake.

Many people are fascinated by snakes, and few are indifferent toward them. Some people have an irrational fear of all snakes, including harmless varieties, whereas others are strongly attracted to them.

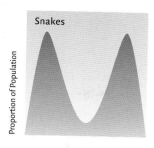

learns to appreciate snakes as valued components of our natural world, laws and attitudes will change, and we will be assured of sharing the country with the snakes and other creatures that are its rightful inhabitants for many, many years to come.

ATTITUDES ABOUT SNAKES

People's feelings about snakes tend to be strongly polarized. They may love them or hate them, but very few people are indifferent to them. It is difficult to explain the powerful love/hate relationship humans have with snakes. One theory says that humans have an innate fear of snakes (ophidiophobia) that is perpetuated by society because a few species are potentially dangerous. The fact that many people today have an appreciation for snakes rather than an aversion clearly demonstrates that the fear can be overridden.

Familiarity is the best way for most individuals to transform their fear of snakes into respect and even admiration. Educating the public about the fascinating behavior, ecological value, and minimal threat associated with snakes is the first step toward developing a general attitude that snakes have far more to offer than most people realize.

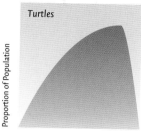

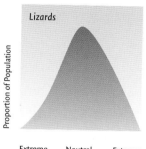

FREQUENTLY ASKED QUESTIONS ABOUT SNAKES

Are snakes slimy? No, snakes are covered with dry scales made of keratin, the same protein that makes up your hair and fingernails. Some snakes are very shiny and thus appear slimy, and some snakes may be wet if they were recently in water.

Is the venom of young snakes more potent than that of adult snakes? The venom of young snakes may differ in composition from the venom of adult snakes because young snakes often feed on different prey than their parents. However, the venom of young snakes is not necessarily more potent, and young venomous snakes cannot inject nearly as much as an adult.

Can snakes other than rattlesnakes make a rattle sound? Yes, many nonvenomous snakes found throughout most of the eastern United States, such as kingsnakes, racers, and ratsnakes, will vibrate their tails when frightened. If they vibrate their tail in dry leaves, it will often sound much like a rattlesnake.

Can snakes bite underwater? Yes, many snakes such as a watersnakes and cottonmouths feed on fish, which they capture with their mouths underwater.

Can you determine the age of a rattlesnake by the number of rattles? No, rattlesnakes add another rattle segment every time they shed their skin. They may shed their skin several times per year. Also, rattle segments may break off from the end of the rattle from time to time.

This eastern green watersnake captured this large catfish underwater and dragged it onto shore to swallow it.

A congregation of timber rattlesnakes on a rocky outcropping

Do nonvenomous snakes breed with venomous ones? No, it is biologically impossible for any venomous snake to successfully breed with a nonvenomous one. However, it is sometimes possible for closely related species to interbreed in nature, although such events are extremely rare.

Do snakes travel in pairs or groups? Snakes are generally not social animals and are usually found alone. However, during the fall or spring breeding seasons, pairs of snakes can sometimes be found together. Some snakes, such as timber rattlesnakes will, congregate at rocky outcroppings to hibernate during the winter.

How do you tell the difference between a venomous snake and a nonvenomous snake? Unfortunately, there is no simple way to distinguish between a venomous snake and a nonvenomous snake. The best possible advice one can follow is that any snake that cannot be positively identified as harmless should not be handled. Many people believe that all snakes with triangular-shaped heads are potentially venomous. Although all pit vipers (rattlesnakes, copperheads, cottonmouths) in the eastern United States have broad triangular heads, so do some nonvenomous species such as watersnakes (genus *Nerodia*). In addition, a nontriangular head does not signify the absence of venom: venomous coral snakes have slender heads with little distinction between head and neck. Another erroneous generalization is that venomous snakes have elliptical pupils, whereas harmless snakes have round pupils; venomous coral snakes also have round pupils. Unfortunately, except for the presence or absence of fangs, no single rule separates all venomous species from all of the harmless ones. The best way to distinguish venomous from nonvenomous snakes is to become familiar with the snakes inhabiting the area in which you live.

Although it appears huge in the picture, this dead eastern diamondback rattlesnake is likely little more than 5 feet in length.

Can the bite from a nonvenomous snake such as a kingsnake or racer make a person sick? The bite from a nonvenomous snake is almost always harmless, and most bites from nonvenomous snakes result in very little pain. However, some people are allergic to the saliva of some snakes, such as the common gartersnake, and can have a reaction if the snake is allowed to chew and break the skin. In most instances, if the bite from a nonvenomous snake is cleaned properly with soap and water, the chance of infection is very low.

Are pictures of giant rattlesnakes real? Yes, many of the pictures are real, but in nearly all cases, the picture is taken in such a way as to make the snake appear much larger than it actually is. Rattlesnakes can get relatively large, but no rattlesnake has ever been scientifically documented to be more than 8 feet long and most "large" rattlesnakes are little more than 5 feet. By holding an object, including a dead rattlesnake, in the foreground in front of a person, it can make the object appear much larger than it actually is. Fishermen perfected the technique of the forced perspective years ago to make their catch look bigger.

what snakes are found where you live?

What kinds of snakes are found in your state?

Occurrence by state of the 63 species of native snakes, listed in the order of those occurring in the most to least number of states.

COMMON NAME	SCIENTIFIC NAME	FL	LA	AL	GA	MS	SC	IL	NC	TN	VA	IN	KY	OH	MD	NJ	DC	PA	WI	WV	DE	MI	NY	MA	CT	RI	VT	NH	ME	TOTAL
Brownsnake	*Storeria dekayi*																													28
Red-Bellied Snake	*Storeria occipitomaculata*																													28
Ring-Necked Snake	*Diadophis punctatus*																													28
Common Gartersnake	*Thamnophis sirtalis*																													28
Eastern Ribbonsnake	*Thamnophis sauritus*																													28
Racer	*Coluber constrictor*																													28
Northern Watersnake	*Nerodia sipedon*																													27
Eastern Hognose Snake	*Heterodon platirhinos*																													27
Milksnake	*Lampropeltis triangulum*																													26
Ratsnake	*Pantherophis obsoletus*																													22
Canebrake/Timber Rattlesnake	*Crotalus horridus*																													22
Eastern Wormsnake	*Carphophis amoenus*																													21
Copperhead	*Agkistrodon contortrix*																													19
Queen Snake	*Regina septemvittata*																													19
Smooth Earthsnake	*Virginia valeriae*																													19
Rough Green Snake	*Opheodrys aestivus*																													18
Common Kingsnake	*Lampropeltis getula*																													16
Smooth Green Snake	*Opheodrys vernalis*																													16
Scarlet Snake	*Cemophora coccinea*																													15
Plain-Bellied Watersnake	*Nerodia erythrogaster*																													14
Corn Snake	*Pantherophis guttatus*																													13
Mole and Prairie Kingsnake	*Lampropeltis calligaster*																													13
Scarlet Kingsnake	*Lampropeltis elapsoides*																													12
Pine Snake	*Pituophis melanoleucus*																													12
Mud Snake	*Farancia abacura*																													11
Cottonmouth	*Agkistrodon piscivorus*																													10
Southeastern Crowned Snake	*Tantilla coronata*																													9
Southern Banded Watersnake	*Nerodia fasciata*																													9
Rough Earthsnake	*Haldea striatula*																													9
Coachwhip	*Masticophis flagellum*																													9
Rainbow Snake	*Farancia erytrogramma*																													9
Pigmy Rattlesnake	*Sistrurus miliarius*																													9
Glossy Crayfish Snake	*Regina rigida*																													8

Common Name	Scientific Name																																Total
Diamondback Watersnake	Nerodia rhombifer	■	■	■	■	■		■		■																								8
Kirtland's Snake	Clonophis kirtlandii	■	■	■	■		■	■	■																									7
Pine Woods Snake	Rhadinaea flavilata	■	■				■	■	■			■		■																				7
Western Ribbonsnake	Thamnophis proximus	■	■	■	■		■	■			■																							7
Western Green Watersnake	Nerodia cyclopion	■	■	■	■		■		■	■																								7
Massasauga	Sistrurus catenatus	■		■	■	■	■	■							■																			7
Eastern Diamondback Rattlesnake	Crotalus adamanteus	■	■			■	■	■				■		■																				7
Coral Snake	Micrurus fulvius	■	■			■	■		■					■																				6
Southern Hognose Snake	Heterodon simus	■	■				■	■	■			■																						6
Brown Watersnake	Nerodia taxispilota	■					■	■	■	■																								5
Fox Snake	Pantherophis vulpinus	■		■	■			■			■																							5
Black Swamp Snake	Seminatrix pygaea	■					■	■	■																									4
Plains Gartersnake	Thamnophis radix	■		■	■			■																										4
Butler's Gartersnake	Thamnophis butleri	■		■	■			■																										4
Eastern Indigo Snake	Drymarchon couperi	■	■				■		■																									4
Salt Marsh Snake	Nerodia clarkii	■	■			■		■																										4
Short-Headed Gartersnake	Thamnophis brachystoma			■	■			■																										3
Western Wormsnake	Carphophis vermis	■		■				■																										3
Bullsnake	Pituophis catenifer	■		■	■																													3
Graham's Crayfish Snake	Regina grahami	■		■	■																													3
Eastern Green Watersnake	Nerodia floridana	■					■	■																										3
Lined Snake	Tropidoclonion lineatum	■		■																														2
Florida Crowned Snake	Tantilla relicta	■						■																										2
Flat-Headed Snake	Tantilla gracilis	■		■																														2
Striped Crayfish Snake	Regina alleni	■						■																										2
Rim Rock Crowned Snake	Tantilla oolitica	■																																1
Short-Tailed Snake	Lampropeltis extenuata	■																																1
Western Hognose Snake	Heterodon nasicus	■																																1
Louisiana Pine Snake	Pituophis ruthveni	■																																1
Great Plains Ratsnake	Pantherophis emoryi	■																																1
Total Native Species		45	42	42	42	39	39	38	34	32	32	25	25	22	21	21	21	20	20	17	17	15	14	14	13	13	12	12	11	11	10	10		

INTRODUCED SPECIES

Common Name	Scientific Name					Total
Brahminy Blind Snake	Ramphotyphlops braminus	■	■	■	■	7
Northern African Python	Python sebae	■				1
Burmese Python	Python molurus bivittatus	■				1
Boa Constrictor	Boa Constrictor	■				1
TOTAL ALL SPECIES		49	43	43	42	42	39	39	38	34	32	32	26	25	25	22	21	21	21	20	20	17	17	15	14	14	13	13	12	12	11	11	10	10

Conservation Status of Snakes in the Eastern United States

Certain species in some states have been recognized for special conservation status for a variety of reasons, as indicated in the key. See species accounts for details of conservation issues. Species not recognized for state or federal conservation concern before 2016 are not included in the table.

■ State conservation concern[1]   ▪ Federal conservation concern[2]

COMMON NAME	SCIENTIFIC NAME	LA	MS	AL	GA	FL	SC	NC	VA	TN	CT	DE	IL	IN	KY	ME	MD	MA	MI	NH	NJ	NY	OH	PA	RI	VT	WV	WI	TOTAL
Canebrake/Timber Rattlesnake	*Crotalus horridus*																												13
Massasauga	*Sistrurus catenatus*																												7
Pine Snake	*Pituophis melanoleucus*																												7
Eastern Ribbonsnake	*Thamnophis sauritus*																												5
Eastern Hognose Snake	*Heterodon platirhinos*																												5
Common Kingsnake	*Lampropeltis getula*																												5
Southern Hognose Snake	*Heterodon simus*																												5
Plain-Bellied Watersnake	*Nerodia erythrogaster*																												4
Racer	*Coluber constrictor*																												4
Queen Snake	*Regina septemvittata*																												4
Smooth Green Snake	*Opheodrys vernalis*																												4
Corn Snake	*Pantherophis guttatus*																												4
Kirtland's Snake	*Clonophis kirtlandii*																												4
Eastern Indigo Snake	*Drymarchon couperi*																												4
Ratsnake	*Pantherophis obsoletus*																												3
Northern Watersnake	*Nerodia sipedon*																												2
Ring-Necked Snake	*Diadophis punctatus*																												2
Copperhead	*Agkistrodon contortrix*																												2
Eastern Wormsnake	*Carphophis amoenus*																												2
Rough Green Snake	*Opheodrys aestivus*																												2
Smooth Earthsnake	*Virginia valeriae*																												2
Scarlet Snake	*Cemophora coccinea*																												2
Mud Snake	*Farancia abacura*																												2
Rainbow Snake	*Farancia erytrogramma*																												2
Coachwhip	*Masticophis flagellum*																												2
Coral Snake	*Micrurus fulvius*																												2
Western Ribbonsnake	*Thamnophis proximus*																												2
Salt Marsh Snake	*Nerodia clarkii*																												2
Fox Snake	*Pantherophis vulpinus*																												2

Common Name	Scientific Name	Total
Butler's Gartersnake	Thamnophis butleri	2
Plains Gartersnake	Thamnophis radix	2
Lined Snake	Tropidoclonion lineatum	2
Pigmy Rattlesnake	Sistrurus miliarius	2
Western Green Watersnake	Nerodia cyclopion	2
Common Gartersnake	Thamnophis sirtalis	1
Mole and Prairie Kingsnake	Lampropeltis calligaster	1
Cottonmouth	Agkistrodon piscivorus	1
Southeastern Crowned Snake	Tantilla coronata	1
Southern Banded Watersnake	Nerodia fasciata	1
Eastern Diamondback Rattlesnake	Crotalus adamanteus	1
Western Wormsnake	Carphophis vermis	1
Bullsnake	Pituophis catenifer	1
Louisiana Pine Snake	Pituophis ruthveni	1
Flat-Headed Snake	Tantilla gracilis	1
Western Hognose Snake	Heterodon nasicus	1
Short-Tailed Snake	Lampropeltis extenuatum	1
Great Plains Ratsnake	Pantherophis emoryi	1
Rim Rock Crowned Snake	Tantilla oolitica	1

TOTAL FOR STATE: 1, 5, 9, 2, 8, 1, 3, 4, 3, 11, 12, 1, 2, 2, 3, 6, 3, 5, 5, 14, 2, 1, 5, 1, 13

NOTES

1. State conservation concern: Species designated to be of state conservation concern, which includes several categories (e.g., endangered, threatened, special concern, protected, species of greatest conservation need, and conservation status under review) whose designations and implications vary among states.

2. Federal conservation concern: Species officially recognized under the U.S. Endangered Species Act, which includes several federal categories (endangered, threatened, proposed as threatened, or under review.) As of 2016, no snakes in the eastern United States have been designated to be in the endangered category.

DISTRIBUTION OF ALL SNAKE SPECIES

The number of native species of snakes in an area varies greatly among and within eastern states. Southern regions generally have a higher species diversity of all reptiles, including snakes. For example, Alabama and Georgia have more than twice as many snake species as Massachusetts and Michigan. How species are distributed within a state depends on a variety of geographic, climatic, and environmental factors as well as historical distribution patterns. As a large state that borders the Mississippi River, Illinois has an inordinately high number of species, including several western species not found in most other eastern states. The color coding on the map indicates approximately how many species are found in different areas.

DISTRIBUTION OF VENOMOUS SNAKE SPECIES

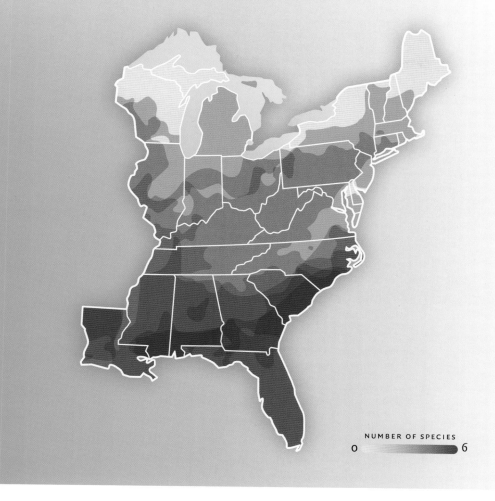

At least one native species of venomous snake occurs in every state. Some regions have as many as six venomous species (e.g., all coastal states from Louisiana to North Carolina) whereas Michigan, Delaware, Rhode Island, New Hampshire, Vermont, and Maine have only one. The color coding on the map indicates how many species are found in different areas.

GLOSSARY

Aestivation   A period of inactivity during dry and/or hot periods in which animals wait for conditions suitable for feeding and other activities to improve.

Albino   An animal completely lacking pigment that provides the color to skin and eyes. Animals lacking only dark pigment, or melanin, are often referred to as "albino" but should more properly be referred to as "amelanistic."

Amphiuma   Any of three species of large, eel-like, aquatic salamanders inhabiting the eastern United States. They have tiny legs and eyes and prefer swampy habitats with aquatic vegetation.

Anal plate   The last scale on the belly of a snake, which covers the cloaca and precedes the scales covering the undersurface of the tail.

Anterior   Referring to the end of the animal toward the head.

Antivenom   A substance administered to a snakebite victim to serve as an antidote or neutralizer of venom injected by a snake. Antivenoms are produced by injecting horses, rabbits, or other animals with venom and then removing their blood to produce a serum. The antivenom commonly used for bites of pit vipers differs from that used for coral snake bites. Also referred to as "antivenin."

Biodiversity   Referring to the numbers, distribution, and abundance of species within a given area.

Bioindicator   A species whose health or condition, either at the individual level or at the population level, is indicative of the condition of the habitat or ecosystem as a whole.

Biomass   The weight of living things in the environment.

Biomonitor   To measure and record the number, types, and characteristics of organisms.

Brumation   A period of inactivity of "cold-blooded" animals, or ectotherms, during cold periods. See also Hibernation.

Cloaca A single opening that serves as the passageway to the outside of an animal for the urinary, digestive, and reproductive tracts.

Clutch A group of eggs laid together at one time by a single individual.

Cold-blooded A nontechnical term that refers to animals whose body temperature is largely determined by environmental conditions and the thermoregulatory behavior of the animal.

Concertina locomotion A method of locomotion used by snakes in which they anchor the posterior portion of their body, extend the anterior portion, and then pull themselves forward.

Courtship The process that precedes mating in snakes and usually involves actions such as rubbing or biting by the male to increase the receptivity of a female to mating.

Diurnal Active during the daytime.

Dorsal Referring to the back of an animal.

Ecdysis The process of shedding the outer layer of skin.

Ecology The study of how organisms interact with their environment.

Ectotherm An animal whose body temperature is largely determined by environmental conditions and the thermoregulatory behavior of the animal.

Endangered Referring to a species or population that is considered at risk of becoming extinct.

Endemic Referring to a species found only in a particular geographic location and nowhere else.

Endotherm An animal that maintains a high body temperature primarily through the use of heat generated by a high metabolic rate.

Extinct Referring to species with no living individuals.

Extirpated Referring to the elimination of a species from a particular region; local extinction.

Family A taxonomic group containing one genus or two or more closely related genera.

Flatwoods Habitat of the lower Coastal Plain of the southern Atlantic and Gulf regions characterized by pines and isolated wetlands in low-lying areas.

Generalist An animal that does not specialize on any particular type of prey or is not restricted to a particular habitat.

Genus (pl. genera) A taxonomic grouping of one species or more than one closely related species.

Hemipenis (pl. hemipenes) One of two penises possessed by individuals of the order Squamata, which includes snakes, lizards, and amphisbaenians.

Hemotoxic Referring to venom that acts by destroying tissues.

Herpetofauna The species of amphibians and reptiles that inhabit a given area.

Herpetologist A scientist who studies snakes, other reptiles, and amphibians.

Hibernation A period of inactivity during cold periods. Also known as "brumation" in reptiles.

Hybrid (v. hybridize) An intermediate form resulting from mating and genetic mixing between individuals of two species.

Incubation period The time period between when eggs are laid and when they hatch.

Intergrade An intermediate form of a species resulting from mating and genetic mixing between individuals of two or more subspecies within a zone where their ranges overlap. Intergrade specimens may possess traits of all subspecies involved.

Jacobson's organs Organs in the roof of the mouth of snakes (and lizards) that detect chemicals from the animal's environment.

Keeled scale A snake scale that has a ridge down the center running parallel to the snake's body. Also known as "rough" scale.

Lateral undulation A method of locomotion used by snakes in which they use waves of the body to push against the ground's surface and move the body forward.

Mimicry A condition in which an animal looks or acts like something else, often a more dangerous animal.

Neurotoxic Referring to venom that acts by disrupting the proper functioning of the nervous system.

Nocturnal Active at night.

Pheromone A chemical that is released by an animal and used as a signal to other animals of the same species. Female snakes will often release pheromones to attract male snakes during the mating season.

Phylogeny The evolutionary relationships among different groups and species of animals.

PIT tag A Passive Integrated Transponder, a glass-encapsulated electronic device injected into the body cavity of animals for identification purposes. PIT tags emit signals that can be read by a PIT tag reader in close proximity.

Pit viper A venomous snake belonging to the family Viperidae and having a heat-sensitive pit between the eye and nostril. The family Viperidae includes the cottonmouth, the copperhead, and rattlesnakes.

Posterior Away from the head of an animal and toward the tail.

Radiotelemetry A method using a radiotransmitter attached to or implanted in an animal to track its movements by locating it using a directional antenna and radio receiver.

Rear-fanged   Referring to the characteristic of some snakes of having enlarged teeth in the back of the mouth.

Rectilinear locomotion   A method of locomotion used by snakes in which they crawl in a straight line and use their ribs and belly scales to push themselves forward. Rectilinear locomotion is generally used by large, slow-moving snakes when they are not in a hurry.

Rookery   A site associated with the development or care of the young.

Sandhills   Habitat in the Coastal Plain characterized by sandy soils, rolling topography, scrub oak, and longleaf or slash pine.

Sidewinding   A method of locomotion used on unstable substrates such as sand by some snakes in which they lift the body off the ground to move forward, leaving a series of unconnected tracks that are parallel to each other but at almost right angles to the direction of movement.

Smooth scale   A snake scale that is completely flat with no ridge down the center.

Specialist   An animal restricted in its choice of diet or habitat.

Species   A species is difficult to define, even for biologists, but a traditional definition has been that a species is an identifiable and distinct group of organisms in which individuals are capable of interbreeding and producing viable offspring under natural conditions.

Subspecies   A taxonomic unit or "race" within a species, usually defined as morphologically distinct and occupying a geographic range that does not overlap with that of other "races" of the species. Subspecies may interbreed naturally in areas of geographic contact (*see* Intergrade).

Taxonomy   The scientific field of classification and naming of organisms.

Ventral   Referring to the belly or underside of an animal.

Warm-blooded   A nontechnical term that refers to an animal that maintains its body temperature primarily through the use of metabolic heat.

FURTHER READING

Ashton, R. E., Jr., and P. S. Ashton. 1981. Handbook of Reptiles and Amphibians of Florida. Part 1: The Snakes. Windward Publishing, Miami, Fla.

Bartlett, R. D., and P. P. Bartlett. 2003. Florida's Snakes: A Guide to Their Identification and Habits. University of Florida Press, Gainesville.

Bartlett, R. D., and P. P. Bartlett. 2006. Guide and Reference to the Snakes of Eastern and Central North America (North of Mexico). University of Florida Press, Gainesville.

Beane, J. C., A. L. Braswell, J. C. Mitchell, W. M. Palmer, and J. R. Harrison III. 2010. Amphibians and Reptiles of the Carolinas and Virginia, 2nd Ed. University of North Carolina Press, Chapel Hill.

Behler, J. L., and F. W. King. 1979. The Audubon Society Field Guide to North American Reptiles and Amphibians. Alfred A. Knopf, New York.

Campbell, J. A., and W. W. Lamar. 2004. The Venomous Reptiles of the Western Hemisphere. 2 vols. Comstock Books in Herpetology. Cornell University Press, Ithaca, N.Y.

Camper, J. D. Forthcoming. The Reptiles of South Carolina. University of South Carolina Press, Columbia.

Carmichael, P., and W. Williams. 2001. Florida's Fabulous Reptiles and Amphibians. World Publications, Tampa, Fla.

Casper, G. S. 1997. Geographic Distributions of the Amphibians and Reptiles of Wisconsin: An Interim Report of the Wisconsin Herpetological Atlas Project. Milwaukee Public Museum, Milwaukee, Wisc.

Conant, R., and J. T. Collins. 1998. A Field Guide to Reptiles and Amphibians of Eastern and Central North America. Third expanded edition. Houghton Mifflin, Boston.

Dorcas, M. E. 2004. A Guide to the Snakes of North Carolina. Davidson College Herpetology Laboratory, Davidson, N.C.

Dorcas, M. E., and J. D. Willson. 2011. Invasive Pythons in the United States: Ecology of an Introduced Predator. University of Georgia Press, Athens.

Dundee, H. A., and D. A. Rossman. 1989. The Amphibians and Reptiles of Louisiana. Louisiana State University Press, Baton Rouge.

Ernst, C. H., and E. M. Ernst. 2003. Snakes of United States and Canada. Smithsonian Books, Washington, D.C.

Ernst, C. H., and G. R. Zug. 1996. Snakes in Question. Smithsonian Institution Press, Washington, D.C.

Gibbons, W. and M. Dorcas. 2015. Snakes of the Southeast. Revised Edition. University of Georgia Press, Athens.

Gibbons, J. W. 1983. Their Blood Runs Cold: Adventures with Reptiles and Amphibians. University of Alabama Press, Tuscaloosa.

Gibbons, J. W., and M. E. Dorcas. 2004. North American Watersnakes: A Natural History. University of Oklahoma Press, Norman.

Gibbons, W., and P. J. West, eds. 1998. Snakes of Georgia and South Carolina. Savannah River Ecology Laboratory Herp Outreach Publication 1, Aiken, S.C.

Gibbs, J. P., A. R. Breisch, P. K. Ducey, and G. Johnson. 2007. The Amphibians and Reptiles of New York State: Identification, Natural History, and Conservation. Oxford University Press, New York.

Greene, H. W. 1997. Snakes: The Evolution of Mystery in Nature. University of California Press, Berkeley.

Harding, J. H. 1997. Amphibians and Reptiles of the Great Lakes Region. University of Michigan Press, Ann Arbor.

Holman, J. A., and J. H. Harding. 2006. Michigan Snakes: A Field Guide and Pocket Reference. Michigan State University Extension, Lansing.

Hunter, M. L., Jr., J. Albright, and J. Arbuckle. 1992. The Amphibians and Reptiles of Maine. Maine Agricultural Experiment Station, Orono.

Jackson, J. J. 1983. Snakes of the Southeastern United States. Cooperative Extension Service. University of Georgia, Athens.

Jensen, J. B., C. D. Camp, W. Gibbons, and M. J. Elliott. 2008. Amphibians and Reptiles of Georgia. University of Georgia Press, Athens.

Krysko, K. L., K. M. Enge, and P. E. Moler. 2011. Atlas of Amphibians and Reptiles in Florida. Final Report, Project Agreement 08013, Florida Fish and Wildlife Conservation Commission, Tallahassee.

Lillywhite, H. B. 2014. How Snakes Work: Structure, Function and Behavior of the World's Snakes. Oxford University Press, New York.

Lohoefener, R., and R. Altig. 1983. Mississippi Herpetology. Mississippi State University Research Center, NSTL Station, Miss.

Martof, B. S., W. M. Palmer, J. R. Bailey, J. R. Harrison III, and J. Dermid.

1980. Amphibians and Reptiles of the Carolinas and Virginia. University of North Carolina Press, Chapel Hill.

Minton, S. A., Jr. 2001. Amphibians and Reptiles of Indiana. Indiana Academy of Science, Indianapolis.

Mitchell, J. C. 1994. The Reptiles of Virginia. Smithsonian Institution Press, Washington, D.C.

Mount, R. H. 1975. The Reptiles and Amphibians of Alabama. Auburn University Agricultural Experiment Station, Auburn, Ala.

Oldfield, B., and J. J. Moriarty. 1994. Amphibians and Reptiles Native to Minnesota. University of Minnesota Press, Minneapolis.

Palmer, W. M., and A. L. Braswell. 1995. Reptiles of North Carolina. University of North Carolina Press, Chapel Hill.

Pinder, M. J., and J. C. Mitchell. 2001. A Guide to the Snakes of Virginia. Wildlife Diversity Special Publication 2. Virginia Department of Game and Inland Fisheries, Richmond.

Powell, R., R. Conant, and J. T. Collins. 2016. Peterson Field Guide to Reptiles and Amphibians of Eastern and Central North America. Fourth edition. Houghton, Mifflin, Harcourt, Boston.

Rossman, D. A., N. B. Ford, and R. A. Seigel. 1996. The Garter Snakes: Evolution and Ecology. Animal Natural History Series, vol. 2. University of Oklahoma Press, Norman.

Rowell, J. C. 2012. The Snakes of Ontario: Natural History, Distribution, and Status. Privately published.

Tennant, A. 2003. Snakes of North America: Eastern and Central Regions. Lone Star Field Guides, Lanham, Md.

Tennant, A., and R. D. Bartlett. 2000. Snakes of North America: Eastern and Central Regions. Gulf Publishing Company, Houston, Tex.

Zim, H. S., and H. M. Smith. 2001. Reptiles and Amphibians. A Golden Guide. St. Martin's Press, New York.

ACKNOWLEDGMENTS

I owe special thanks to several people without whose help I would have been many more months completing this book. First and foremost, I owe a debt of inexpressible gratitude to Michael E. Dorcas for the countless hours he spent on all facets of this book. His hands-on experience, especially with all eastern snakes including the introduced pythons in Florida, was invaluable. He contributed extensively to the writing of species accounts and general sections, to the preparation of tables, and to the cataloguing, selecting, and handling of photographic files as well as to communicating with photographers. His careful attention to details associated with the numerous geographic ranges for all species was vital in the preparation of the maps.

Margaret Wead's constant attention to details in the preparation of photographic slides and digital images, copying of text and drawings, and acquiring library materials was invaluable. Katie Greene did an outstanding job of organizing and cataloging thousands of images, assisting with the preparation of geographic range maps, tracking down publications, and preparing charts and tables for the book. Susan Lane Harris and Nicholas Harris prepared the indexes.

Kraig Adler, Cornell University, did an excellent job reviewing the etymology sections for every species and offering helpful suggestions. I am greatly appreciative of his willingness to help with this task in a timely fashion. In addition to Kraig, I thank Phyllis Britt, Xavier Glaudas, Ann Steeper, and Margaret Wead for language translations.

I also appreciate the assistance of several individuals who have helped over the years with particular organizational issues on previous projects that proved helpful later during preparations for this book. Notable among these are Teresa Carroll, Sarah Collie, Jackie Guzy, Shannon Pittman, Steven J. Price, Meagan Thomas, and Patricia West. John Jensen of the Georgia Department of Natural Resources provided helpful suggestions, information, and guidance on the biology of several southeastern snakes. Jeff Camper was extremely generous in allowing me access to his book *Reptiles of South Carolina* prior to its publication, which provided information about the biology and distribution patterns of several species of eastern snakes. I owe special thanks to Joe Mitchell for reading and providing useful comments on the entire book.

I am very grateful to the many herpetologists who provided a spectacular array of images for use in the book. The magnificent color images generously offered by so many experts have added immensely its value as a practical field guide for identification of eastern snakes. The following individuals provided images to examine: Michael Barron, Dick Bartlett, Jeffrey C. Beane, James Beasley, Ted Braun, Carl R. Brune, Evin T. Carter, Jonathan Cooley, James R. Craig, Jacob A. Daly, Jeffrey G. Davis, Ian Deery, Jessika Dorcas, Andrew M. Durso, Jason R. Folt, Mike Gibbons, Parker Gibbons, Christopher Gillette, Matt Greene, Greg C. Greer, Iwo Gross, Eitan Grunwald, Ron Grunwald, Jackie Guzy, James Harding, Aubrey M. Heupel, Jennifer G. High, Pierson Hill, T. J. Hilliard, Andrew Hoffman, Cary Howe, John B. Jensen, Wade G. Kalinowsky, Robert A. Kennamer, Kris Leefers, Thomas M. Luhring, Reagan Lunn, Mike D. Martin, Chris McEwen, Kevin R. Messenger, Lori Oberhofer, Lance Paden, Todd W. Pierson, Mike Pingleton, Shannon Pittman, Kory G. Roberts, Michael R. Rochford, Ron Rozar, Michael Sankewitsch, Nick Scobel, Nathan Shepard, Rhett Stanberry, Dirk J. Stevenson, Kevin Stohlgren, Daniel Thompson, Brian D. Todd, Robert Wayne Van Devender, Phil Vogrinc, Margaret Wead, John White, Rose Williams, J. D. Willson, and Robert T. Zappalorti. Meg Francoeur, Julian Lockwood, Lori Oberhofer (National Park Service), Dirk Stevenson, Kelsey L. Turner, John White (Virginia Herpetological Society), and J. D. Willson (University of Arkansas) provided assistance in acquiring particular images.

I turned to herpetological colleagues for their expert knowledge of the distribution and abundance of snakes in particular states or regions. Although I hold ultimate responsibility if any errors in the geographic range maps exist, suggestions and recommendations by these individuals aided greatly in estimating past and current distribution patterns of eastern snakes. For their help I thank Chuck Ancelli, Jim Andrews, Jeff Beane, Jeff Camper, Alvin Breisch, Bill Brown, Carl Brune, Jeff Camper, Gary Casper, Jeff Davis, Phillip deMaynadier, Mark P. DesMeules, Jim Harding, Mac Hunter, Mike Klemens, Bruce Kingsbury, Greg Lipps, John MacGregor, Michael Marchand, Walt Meshaka, David Misfud, Joe Mitchell, Tom Pauly, Trevor Persons, Chris Phillips, Chris Raithel, Howard Reinert, Jesse Rothacker, Chuck Smith, Tom Tyning, Doug Wynn, and Robert Zappalorti.

Numerous associates whose expertise on the natural history of particular species of snakes aided greatly in making each species account as thorough and up-to-date as possible. I thank the following for providing comments on various species accounts and topical sections: Kimberly Andrews, Kraig Adler,

Dick Bartlett, Rick Bauer, Jeff Beane, Steve Bennett, Al Breisch, Kurt Buhlmann, Carl Brune, John Byrd, Carlos Camp, Gary Casper, Vince Cobb, Jeff Davis, Brett DeGregorio, Will Dillman, Terry Farrell, Darren Fraser, Chris Gillette, Xavier Glaudas, Steve Godley, Jim Godwin, Sean Graham, Judy Greene, Katie Greene, Greg Greer, Jackie Guzy, Kerry Hansknecht, Jim Harding, Susan Harris, Natalie Haydt, Pierson Hill, Bob Jaeger, John Jensen, Emma Johnson, Bruce Kingsbury, Yale Leiden, Joe Letsche, Greg Lipps, Jeff Lovich, Tom Luhring, John MacGregor, Walt Meshaka, Brian Metts, David Mifsud, Mark Mills, Tony Mills, Joe Mitchell, Paul Moler, Brad Moon, Lance Paden, Emma Rose Parker, Tom Pauley, Scott Pfaff, Chris Phillips, Melissa Pilgrim, Shannon Pittman, Mike Plummer, Sean Poppy, Steve Price, Bob Reed, David Scott, Floyd Scott, Dirk Stevenson, Brian Todd, Stan Trauth, Tracey Tuberville, Anna Tutterow, Tom Tyning, Wayne VanDevender, James VanDyke, Phil Vogrinc, Jayme Waldron, Mark Waters, Margaret Wead, J. D. Willson, Chris Winne, Doug Wynn, Cameron Young, and Robert Zappalorti.

I thank the following for providing lists of the most commonly encountered species of snakes in suburban areas of large cities of the eastern United States: Al Breisch, Joe Butler, John Byrd, Gary Casper, Jeff Davis, Mike Dloogatch, Terry Farrell, Lisa Rania Ganser, Greg Greer, Jim Harding, John Jensen, Bruce Kingsbury, Walt Meshaka, Joe Mitchell, Paul Moler, Chris Phillips, Bob Thomas, Tom Tyning, Patricia J. West, J. D. Willson, and Robert Zappalorti.

I also appreciate the assistance provided by numerous friends and colleagues in responding to specific requests regarding a variety of topics, including geographic range distributions, problematic locality records, journal articles, and reported observations of unusual snake behavior. I particularly thank the following individuals for their help in these categories: Al Breisch, Ron Brenneman, Gary Casper, Heyward Clamp, Ted Clamp, Emile DeVito, Mike Dloogatch, Carl Ernst, John Jensen, Stephanie Lockwood, Rudy Mancke, John Macgregor, Tony Mills, Scott Pfaff, Chris Phillips, Robert Powell, David Steen, Travis Taggart, and Robert Zappalorti. Melissa Buchanan, Design and Production Manager at the University of Georgia Press, did an outstanding job coordinating editing of several iterations of manuscripts and proofs. Mindy Hill did a superb job of the overall book design. Because this book supports the efforts of Partners in Amphibian and Reptile Conservation (PARC) to promote education about reptiles and amphibians, I thank the many PARC members who offered encouragement, advice, and enthusiastic support.

Finally, for their personal support and understanding during the preparation of this book I thank Tammy, Taylor, Jessika, and Zachary Dorcas; Carolyn Gibbons; Laura Gibbons and Ron Curtis; Jennifer, Jim, and Sam High; Susan Lane, Keith, and Nicholas Harris; and Michael, Jennifer, Allison, and Parker Gibbons.

PHOTO CREDITS

Thank you to the following individuals and organizations for providing photographs:

Mike Barron
Photograph on page 353 (top).

Dick Bartlett
Photographs on pages 105 (bottom), 181, 268 (both).

Jeffrey C. Beane
Photographs on pages 4 (top left), 6 (middle), 27 (right), 44, 133, 137, 148 (top), 157 (bottom), 159, 220 (bottom), 234 (left), 236, 267 (top), 293, 296, 298, 382.

James Beasley
Photographs on page 20 (all).

Tim Borski
Photograph on page 91.

Ted Braun
Photograph on page 15 (left).

Carl R. Brune
Photographs on pages 7 (top), 9 (top and middle left), 18 (top), 54 (both), 56, 61 (bottom), 64, 79, 93 (bottom), 136, 165, 177, 194, 212, 301 (bottom), 303 (bottom), 372 (bottom).

Evin T. Carter
Photographs on pages 81 (bottom), 320.

Jacob A. Daly
Photographs on pages 88 (right), 264, 309.

Jeffrey G. Davis
Photographs on pages 73, 121 (top), 193 (top right), 376 (top).

Ian Deery
Photographs on pages ii, 3 (bottom), 28 (top), 51, 199, 327, 381 (bottom).

Michael Dorcas
Photographs on pages 299 (right), 343 (right), 344, 367.

Jessika Dorcas
Photograph on page 299 (left).

Andrew M. Durso
Photographs on pages 60, 153, 193 (bottom), 203, 207, 240, 288, 313.

Jason R. Folt
Photographs on pages 17 (middle right), 134, 140, 190, 211, 250, 251 (top left), 253 (bottom), 321, 322, 372 (top).

Mike Gibbons
Photographs on pages 369 (bottom), 386 (right).

Parker Gibbons
Photograph on page 3 (top).

Whit Gibbons
Photographs on pages 21 (middle left and right), 40 (bottom), 369 (top), 370, 371 (top), 376 (bottom right), 377, 379 (both), 383, 386 (left).

Christopher Gillette
Photographs on pages 14 (bottom), 21 (top), 92, 94 (bottom), 102, 103, 105 (top), 132 (top), 170 (bottom), 201 (left), 225 (bottom), 228, 317, 341, 346 (top), 352 (top).

Matt Greene
Photographs on pages 34 (left), 42, 260, 332 (left), 384 (top).

Iwo Gross
Photographs on pages 23 (middle), 332 (right), 371 (bottom).

Eitan Grunwald
Photographs on pages 15 (right), 69, 148 (bottom), 163 (bottom), 167, 202 (bottom), 256 (top left and bottom), 326, 335, 389.

Ron Grunwald
Photographs on pages 10 (top), 30, 38, 84, 141, 271, 288.

Aubrey M. Heupel
Photographs on pages v, 6 (top), 10 (bottom right), 25 (top left, bottom right), 29 (bottom), 116, 120, 131 (top), 139 (bottom), 147 (top), 152, 168 (bottom), 196 (bottom), 219, 238, 261 (right), 285 (bottom), 385.

Jennifer G. High
Photograph on page 380.

Pierson Hill
Photographs on pages 5 (bottom), 24 (top left), 41 (bottom), 100, 150, 151, 175 (all), 176, 188, 205, 208, 210 (right), 221, 225 (top and middle), 249, 287, 360, 361.

TJ Hilliard
Photographs on pages 229, 251 (bottom), 308, 381 (top).

Andrew Hoffman
Photographs on pages 16, 32 (top), 40 (top), 62 (top), 65 (bottom), 72, 96, 98 (top), 107, 108 (top), 144, 160 (top), 163 (top), 204, 210 (left), 253 (top left), 275, 303 (top), 324.

Cary Howe
Photographs on pages 139 (top), 173, 197, 218, 220 (top), 295 (both).

John B. Jensen
Photographs on pages 36, 266.

Robert A. Kennamer
Photograph on page 368.

Kris Leefers
Photograph on page 351 (bottom).

Reagan Lunn
Photograph on page 43.

Mike D. Martin
Photographs on pages 7 (bottom), 9 (top right), 67, 68, 155 (top), 246, 297 (top), 330, 333, 340, 384 (bottom).

Kevin R. Messenger
Photographs on pages i, 35 (top), 127, 142 (top), 242 (bottom), 262, 263 (bottom), 328.

Lori Oberhofer
Photographs on pages 350, 351 (middle), 352 (bottom), 355.

Lance Paden
Photographs on pages 17 (bottom), 22, 25 (bottom left), 132 (bottom), 171, 201 (right), 258 (top), 338.

Todd W. Pierson
Photographs on pages 59, 123, 126, 161, 162, 191, 193 (top middle), 214 (top), 281, 311, 346 (bottom), 354.

Mike Pingleton
Photograph on page 189.

Kory G. Roberts
Photographs on pages 70 (both), 174, 206, 259, 273, 276, 282.

Michael R. Rochford
Photographs on page 353 (bottom).

Ron Rozar
Photographs on pages 337 (bottom), 356, 357, 358, 362.

Michael Sankewitsch
Photograph on pages 118 (bottom).

Nick Scobel
Photograph on pages 182 (top), 183, 318.

Nathan Shepard
Photographs on pages 31, 34 (right), 35 (bottom), 57 (bottom), 61 (top), 74, 75, 82 (top), 89, 110 (top), 118 (top), 129 (middle), 154, 160 (bottom), 195, 237, 245, 253 (top right), 255 (bottom), 263 (top), 267 (bottom), 306, 312, 315 (top left), 378.

Rhett Stanberry
Photographs on pages 129 (top), 307, 339.

Kevin Stohlgren
Photographs on pages 17 (top), 26 (left), 27 (left), 41 (top), 93 (top), 109, 111, 128, 145 (top), 147 (bottom), 155 (bottom), 192, 196 (top), 226, 227, 231, 269, 289, 290 (top), 301 (top), 315 (top right and bottom), 323 (right), 329, 337 (top), 345 (both).

Daniel Thompson
Photographs on pages 7 (middle), 23 (bottom), 57 (top), 108 (bottom), 149 (right), 157 (top), 158, 223 (top), 235, 252, 277, 304, 316.

Brian D. Todd
Photographs on pages 24 (top right), 110 (bottom), 115 (bottom), 145 (bottom), 285 (top).

Robert Wayne Van Devender
Photographs on pages 5 (top), 76, 77, 88 (left), 213, 215 (top).

Margaret Wead
Photograph on page 368.

John White
Photographs on pages 17 (middle left), 18 (bottom), 19 (bottom), 25 (top right), 28 (bottom), 37 (bottom), 62 (bottom), 81 (top), 121 (bottom), 142 (bottom), 149 (left), 170 (top), 179, 214 (bottom), 290 (bottom), 310, 323 (left), 376 (bottom left).

Rose Williams
Photograph on page 291.

J. D. Willson
Photographs on pages 4 (top right), 10 (bottom left), 19 (top), 37 (top), 53, 65 (top), 79 (top), 82 (bottom), 83, 86, 98 (bottom), 106, 113, 124, 129 (bottom), 168 (top), 193 (top left), 195, 215 (bottom), 217, 232, 234 (right), 239, 242 (top), 244, 248, 251 (top right), 255 (top), 256 (top right), 258 (bottom), 261 (left and middle), 265, 270, 271, 278, 279, 284, 343 (left), 349, 351 (top), 388.

Robert T. Zappalorti
Photographs on pages 5 (middle), 14 (top), 29 (top), 32 (bottom), 99, 117, 131 (bottom), 166, 180, 182 (bottom), 184, 186, 200, 202 (top), 216, 223 (bottom), 300.

INDEX OF SCIENTIFIC NAMES

Pages references in **bold** refer to species accounts. Page references in *italics* refer to illustrations not contained within a species account.

Agkistrodon contortrix, 293, **301–6**
Agkistrodon piscivorus, **307–12**

Boa constrictor, 344, **360–63**
Boa contortrix, 306

Carphophis amoenus, **79–82**
Carphophis vermis, **83–85**
Cemophora coccinea, **113–16**
Clonophis kirtlandii, **72–75**
Coluber, 222
Coluber aestivus, 120
Coluber amoenus, 82
Coluber catenifer, 191
Coluber coccinea, 116
Coluber constrictor, **212–18**
Coluber couperi, 228
Coluber erythrogaster, 268
Coluber erytrogrammus, 292
Coluber fasciatus, 259
Coluber fulvius, 340
Coluber getulus, 178
Coluber melanoleucus, 183
Coluber obsoletus, 198
Coluber punctatus, 110
Coluber saurita, 140
Coluber Sebae, 359
Coluber simus, 157
Coluber sipedon, 254
Coluber sirtalis, 132
Coluber vernalis, 123

Contia pygaea, 234
Crotalinus catenatus, 321
Crotalus adamanteus, **329–34**
Crotalus horridus, **322–28,** *328*
Crotalus miliarius, 317

Diadophis punctatus, **106–10**
Dromicus flavilatus, 104
Drymarchon couperi, **223–28**

Elaphe, 198
Elapidae, 298
Eryx braminus, 348
Eutaenia brachystoma, 78
Eutaenia sackenii, 140

Farancia abacura, **284–88**
Farancia erytrogramma, **289–92**

Haldea striatula, **57–60**
Heterodon nasicus, **150–53**
Heterodon platirhinos, **145–49**
Heterodon simus, **154–57**

Indotyphlops braminus, 348

Lampropeltis calligaster, **158–61**
Lampropeltis elapsoides, 166, **167–70**
Lampropeltis extenuata, **99–101**
Lampropeltis getula, **173–78**
Lampropeltis triangulum, **162–66,** *167, 170*

Liochlorophis, 123
Liodytes, 234, 244

Masticophis flagellum, **219–22**
Micrurus fulvius, 170, **335–40**
Micrurus tener, 340

Natrix, 75
Nerodia clarkii, **260–63**
Nerodia cyclopion, **281–83**
Nerodia erythrogaster, **264–68**
Nerodia fasciata, **255–59**
Nerodia floridana, **277–80**
Nerodia rhombifer, **273–76**
Nerodia sipedon, 229, **249–54**
Nerodia taxispilota, **269–72**

Opheodrys aestivus, **117–20**
Opheodrys vernalis, **121–23**

Pantherophis emoryi, 199, **204–7**
Pantherophis guttatus, **199–203**, 207
Pantherophis obsoletus, 171, **192–98**
Pantherophis slowinskii, 203, 207
Pantherophis vulpinus, **208–11**
Pituophis catenifer, 185, 187, **188–91**
Pituophis melanoleucus, **179–83**, 187
Pituophis ruthveni, **184–87**
Python bivittatus, 355
Python molurus bivittatus, 341, 344, **349–55**
Python sebae, 344, **356–59**

Ramphotyphlops braminus, 344, **345–48**
Regina alleni, **242–44**
Regina clarkii, 263

Regina grahamii, **239–41**
Regina kirtlandii, 75
Regina rigida, **245–48**
Regina septemvittata, **235–38**
Rhadinaea, 105
Rhadinaea flavilata, **102–5**

Scotophis emoryi, 207
Scotophis vulpinus, 211
Seminatrix, 234
Seminatrix pygaea, **231–34**
Sistrurus catenatus, **318–21**
Sistrurus miliarius, **313–17**
Stilosoma, 101
Stilosoma extenuatum, 101
Storeria dekayi, **61–64**
Storeria occipitomaculata, **65–68**
Storeria victa, 64

Tantilla coronata, **86–88**, 94
Tantilla gracilis, **96–98**
Tantilla oolitica, **89–91**
Tantilla relicta, **92–95**
Thamnophis brachystoma, **76–78**, 136
Thamnophis butleri, 78, **134–36**
Thamnophis ordinoides, 144
Thamnophis proximus, **141–44**
Thamnophis radix, **124–27**, 136
Thamnophis sauritus, **137–40**, 144
Thamnophis sirtalis, 111, **128–33**
Tropidoclonion lineatum, **69–71**
Tropidonotus rhombifer, 276

Virginia, 71
Virginia valeriae, **53–56**, 57

INDEX OF COMMON NAMES

Pages references in **bold** refer to species accounts. Page references in *italics* refer to illustrations not contained within a species account.

African python, 359
African rock python, 359
Asian python, 355
Asian rock python, 355

banded watersnake, 31, 373
banded watersnake, southern, **255–59**
black runner, 218
black snake, 11, 198
black swamp snake, **231–34**
black swamp snake, Carolina, 231
black swamp snake, north Florida, 231
black swamp snake, south Florida, 231
blind snake, 348
blotched watersnake, 264, 266, 268
blue bullsnake, 228
blue gopher snake, 228
blue runner, 218
boa constrictor, 9, 344, **360–63**
Brahminy blind snake, 9, 344, **345–48**
broad-banded watersnake, 255, 259
brownsnake, 28, **61–64**, *376*
brown watersnake, 14, 36, **269–72**, 274
bullsnake, 185, 187, **188–91**
Burmese python, **349–55**, 356; defensive behavior of, 359; as introduced species, 9, 344, 348; as predator, 358
bushmaster, 296

candy cane snake, 340
canebrake rattlesnake, 25, 33, 44, **322–28**, *385*

Carolina watersnake, 249, 251–52, 253, 254
chicken snake, 11, 198
coachwhip, 21, 22, 35, 38, **219–22**
coachwhip, eastern, 219
cobra, 298, 344
copperbelly watersnake, 264, 266, 268, 383, 385
copperhead, 9, 29, 32, 40, **301–6**, *308*; bites by, 298; daily activity of, 34; description of, 40, 388; distribution of, 44, 372, 373; feeding behavior of, 16; hibernation of, 33; lifespan of, 226; predation of, 165, 177, 216; as predator, 21, 197; venom of, 296
coral snake, 9, 40, **335–40**; bites by, 298; defensive behavior of, 24; description of, 40–41, 42, 170, 388; habitat of, 44; mimicry of, 113; as predator, 21, 56, 59, 88, 100, 115; venom of, 19, 298
coral snake, eastern, 165, 169, 335, 340
coral snake, Texas, 335, 340
corn snake, 30, **199–203**, 379, 381; daily activity of, 34; description of, 41, 204, 207
corn snake, Slowinski's, 203, 207
cottonmouth, 9, 42, **307–12**, *382*; bites by, 298; defensive behavior of, 23, 24, 295–96; description of, 40, 42; feeding behavior of, 16, 274, 387; habitat of, 11; hibernation of, 32, 33; and male-male combat, 26; predation of,

cottonmouth (continued)
20, 177; as predator, 139, 144, 238, 244, 258, 267, 272, 274, 282, 287; venom of, 296
cottonmouth, eastern, 307
cottonmouth, Florida, 312
cottonmouth, western, 307, 312
crayfish snake, glossy, 19, 37, **245–48,** 384
crayfish snake, Graham's, 28, **239–41**
crayfish snake, striped, 19, 41, **242–44**
crowned snake, 19
crowned snake, Florida, 91, **92–95,** 100
crowned snake, rim rock, **89–91**
crowned snake, southeastern, **86–88,** 91

DeKay's snake, 64, 372
diamondback rattlesnake, eastern, 14, 23, 44, **329–34,** 383
diamondback watersnake, 15, 271, **273–76**

earthsnake, 38, 373
earthsnake, rough, 41, **57–60**
earthsnake, smooth, 7, 41, **53–56,** 57
Egyptian cobra, 344
Emory's ratsnake, 207

fer-de-lance, 296
flat-headed snake, **96–98**
Florida brownsnake, 64
Florida green watersnake, 280
Florida watersnake, 255, 256
flowerpot snake, 348
fox snake, **208–11**
fox snake, eastern, 208–11
fox snake, western, 208–11

gartersnake, 25, 33; description of, 37, 42; distribution of, 372; musk of, 23; as predator, 64; and reproduction, 27
gartersnake, blue-striped, 130, 133
gartersnake, Butler's, 44, **134–36**
gartersnake, Chicago, 130, 133
gartersnake, common, 6, 25, **128–33,** 134, 373

gartersnake, eastern, 10, 128, 130, 133
gartersnake, maritime, 130–31, 133
gartersnake, Plains, **124–27,** 136
gartersnake, San Francisco, 132
gartersnake, short-headed, 5, **76–78,** 136
glossy crayfish snake, 19, 37, **245–48,** 384
Graham's crayfish snake, 28, **239–41**
green snake, 37
green snake, rough, 4, **117–20,** 121, 122, 373
green snake, smooth, 4, **121–23**
green watersnake, eastern, **277–80**

harlequin snake, 340
hognose snake, 16, 24, 34, 42
hognose snake, eastern, 3, 19, 24, 34, 37, 42, **145–49**
hognose snake, southern, 37, 44, **154–57,** 381, 383
hognose snake, western, **150–53**
hoop snake, 288

Indian python, 355
indigo snake: decline of, 383; description of, 38, 41; as endangered, 262, 385; as predator, 18, 21, 197, 316, 326, 332, 338
indigo snake, eastern, 22, 25, 26, 29, **223–28**

king cobra, 344
kingsnake: and male-male combat, 26; mimicking rattlesnake, 41, 387; as pet, 379, 380; as predator, 16, 24, 56, 59, 64, 68, 74, 88, 103, 139, 144, 161, 165, 197, 202, 233, 238, 244, 247, 267, 280, 282, 305, 311, 316, 326, 332, 338
kingsnake, Apalachicola, 175, 178
kingsnake, black. *See* kingsnake, eastern black
kingsnake, chain, 178
kingsnake, common, 21, **173–78**
kingsnake, eastern, 18, 173, 174, 175
kingsnake, eastern black, 173, 174, 178
kingsnake, Florida, 173, 175, 178

414 *Index of Common Names*

kingsnake, mole, **158–61**
kingsnake, Outer Banks, 175, 178
kingsnake, Prairie, **158–61**
kingsnake, short-tailed, 101
kingsnake, speckled, 173, 175, 178
Kirtland's snake, **72–75,** 372
krait, 203, 298

Lake Erie watersnake, 249, 251, 252, 254
lined snake, **69–71**
Louisiana pine snake, **184–87**

mamba, 298
massasauga, 35, 44, 152, 153, 317, **318–21**
massasauga, eastern, 319
midland watersnake, 249, 251, 252, 254
milksnake, 126–27, **162–66,** 170
milksnake, eastern, 162–65, 167
milksnake, Louisiana, 42, 162, 164, 165, 166
milksnake, red, 163, 164, 166, 167
Mississippi green watersnake, **281–83**
moccasin. See cottonmouth
mud snake, 24, 25, 28, 29, **284–88**; tail spine of, 16, 42, 78, 292
mud snake, eastern, 284
mud snake, western, 284

northern African python, 344, 348, **356–59**
northern African rock python, 359
northern watersnake, 17, 18, 44, **249–54,** 376; description of, 265; distribution of, 373; and reproduction, 28

pigmy rattlesnake, 297, **313–17,** 384; distribution of, 44, 373; juveniles of, 16; predation of, 216
pigmy rattlesnake, Carolina, 313, 314
pigmy rattlesnake, dusky, 313, 314, 317
pigmy rattlesnake, western, 313, 317
pine snake, 32, 33–34, 38, **179–83**; defensive behavior of, 24, 41, 191; habitat of, 44; and reproduction, 29
pine snake, black, 179, 180, 181, 183, 262

pine snake, Florida, 179, 180, 183
pine snake, Louisiana, **184–87**
pine snake, northern, 179, 184, 185
pine woods snake, **102–5**
plain-bellied watersnake, **264–68**
python, Burmese. See Burmese python
python, Indian, 355
python, northern African, 344, 348, **356–59**
python, northern African rock, 9
python, reticulated, 344

queen snake, 19, **235–38,** 373

racer, 7, 14, 27, 43, **212–18**; bites by, 26; distribution of, 16; as diurnal, 33–34; juveniles of, 33, 38; mimicking rattlesnake, 41, 387; as predator, 21, 56, 59, 64, 68, 70, 103, 165, 197, 238, 316, 320, 338; sidewinder motion of, 31; as state reptile, 133
racer, black, 31
racer, blackmask, 212, 213, 218
racer, blue, 212, 214, 217
racer, brown-chinned, 212, 216, 218
racer, buttermilk, 212, 214, 215, 218
racer, eastern yellowbelly, 218
racer, Everglades, 212, 215, 218
racer, northern black, 212
racer, southern black, 212, 218
racer, tan, 214, 217–18
rainbow snake, 288, **289–92**; description of, 37; eggs of, 29; tail spine of, 42, 78
rainbow snake, south Florida, 292
ratsnake, **192–98,** 377; bites by, 26; concertina motion of, 31; distribution of, 372, 373; hibernation of, 33; juveniles of, 33, 38; lifespan of, 226; and male-male combat, 26; mimicking rattlesnake, 387; musk of, 23, 202; as pet, 380; and rodent control, 381; scales of, 41
ratsnake, black, 192–93, 194
ratsnake, eastern, 198
ratsnake, Everglades, 192–93, 198
ratsnake, gray, 192–93, 195, 198

Index of Common Names 415

ratsnake, Great Plains, 199, **204–7**
ratsnake, greenish, 198, 386
ratsnake, red, 203
ratsnake, Texas, 192–93, 198
ratsnake, western, 198
ratsnake, yellow, 192–93, 195, 198
rattlesnake, 9; distinctive characteristics of, 40, 42; distribution of, 376; fangs of, 297; hibernation of, 33; and male-male combat, 26; mimicry of, 191, 216; predation of, 165, 177; rattles of, 387; and rectilinear motion, 31; tail length of, 39; venom of, 296, 298. See also canebrake rattlesnake; diamondback rattlesnake, eastern; massasauga; pigmy rattlesnake; timber rattlesnake
red-bellied snake, 5, 24, 37, **65–68**
red-bellied watersnake, 264, 265–66, 267
red-tailed boa, 363
reticulated python, 344
ribbonsnake, 38, 39
ribbonsnake, blue-striped, 133, 137, 140
ribbonsnake, eastern, 10, **137–40**
ribbonsnake, Gulf Coast, 141, 142, 144
ribbonsnake, northern, 137, 140
ribbonsnake, peninsula, 137, 140
ribbonsnake, western, **141–44**
ring-necked snake, **106–10,** 386; defensive behavior of, 24; distribution of, 373; habitat of, 44; musk of, 23; predation of, 84
rock python, 359

salt marsh snake, 11, **260–63**
salt marsh snake, Atlantic, 260, 262, 263; as endangered, 385

salt marsh snake, Gulf, 260
salt marsh snake, mangrove, 260, 263, 263
scarlet kingsnake, 3, **167–70,** 381; and milksnake, 162–64, 166; and mimicry, 42, 339
scarlet snake, 34, 35, **113–16**; distribution of, 373; and mimicry, 42, 339
short-tailed snake, 5, **99–101**
sidewinder rattlesnake, 31
Slowinski's corn snake, 203, 207
southern banded watersnake, 261
striped crayfish snake, 19, 41, **242–44**
"super snake," 359

timber rattlesnake, 23, **322–28**; and basking, 34; extinction of in Maine, 60; hibernation of, 387; lifespan of, 226; as state reptile, 133

water moccasin. See cottonmouth
watersnake, 27, 34; bites by, 26; defensive behavior of, 23; description of, 41, 42, 388; feeding behavior of, 387; as habitat indicator, 381; habitat of, 44; hibernation of, 32; predation of, 310. See also individual species
western green watersnake, **281–83**
whipsnake, 222
wormsnake, 38, 348
wormsnake, eastern, 78, **79–82**
wormsnake, western, 78, **83–85**

yellow-bellied watersnake, 264, 266, 268